THE
GENIUS
WITHIN

Also by David Adam

The Man Who Couldn't Stop

THE GENIUS WITHIN

Unlocking Your Brain's Potential

DAVID ADAM

PEGASUS BOOKS
NEW YORK LONDON

THE GENIUS WITHIN

Pegasus Books Ltd
148 West 37th Street, 13th Floor
New York, NY 10018

First Pegasus Books hardcover edition February 2018

ISBN: 978-1-68177-674-3

10 9 8 7 6 5 4 3 2 1

Printed in the United States of America
Distributed by W. W. Norton & Company, Inc.

The idea was quite logical; a parasite and landowner naturally supposed that intelligence was a marketable commodity like everything else, and that in Switzerland especially it could be bought for money. The case was entrusted to a celebrated Swiss professor, and cost thousands of roubles; the treatment lasted five years. Needless to say, the idiot did not become intelligent, but it is alleged that he grew into something more or less resembling a man.

Fyodor Dostoyevsky, The Idiot

CONTENTS

INTRODUCTION

Your brain is capable of so much more. I know this because, although your brain is unique, it is nothing special. There are billions of brains just like yours. And in some of those brains – the ones just like yours – something extraordinary happens.

Something extraordinary happened to my brain, and that's why I am writing this book. The change opened my eyes to what is possible. My concentration improved, my memory sharpened, my cognitive skills expanded. I became a more fluent communicator and a more sympathetic listener. My productivity at work soared. My home life became happier and more content. And I did it all by finding and activating a part of my brain that had lain dormant for too long.

This part of my brain probably can't be pointed to on a scan. It's more a function of my mind, a door within my consciousness to which I was handed a key. For, despite what you might have heard, it's not true that we only use 10 per cent or so of our brain, which leaves the rest untapped and rich with potential. Our brain cells are overloaded with work, so much so that most have several jobs to do. None are idle.

But it's true that we only access a fraction of what that brain – your brain – could do. Most call it the mind, but you can name it spirit, awareness, consciousness, or the ghost in the machine, whatever term you like. What matters is that it can be altered. It is not much of the brain's *structure* that lies unused, but much of the brain's – your brain's – *function*.

Mapping and understanding brain function and how it can be changed is a frontier of modern neuroscience, the defining discipline of this twenty-first century. And it comes down to connections. Just as the ancients imposed patterns and pictures onto the randomness of the stars, so the brain relies on circuits, sequences and constellations of activity to produce co-ordination and cognition from its billions of individual cells. From memories and mathematics to grief, insight and genius, all of it is formed from the way brain cells make and break links with their neighbours, and how they use these links to communicate. And here's the kicker: science now has the tools to manipulate *and to strengthen* those links on demand. Modern brain science is not just about observing any more. It can intervene, to change the way the brain and the mind works. To make it work better.

My brain was made to work better after I received therapy for mental illness. I had severe obsessive-compulsive disorder (OCD) that showed itself as a wildly exaggerated and irrational fear of HIV and Aids. I had a blind spot in my mental functioning that could not accept very small risks – but only as applied to this single disease. My treatment was cognitive-behavioural therapy, and through a series of mental exercises I learned to deal with and move on from what had previously been debilitating anxieties over, for example, a ridiculous obsessive thought that contaminated blood had fallen into my eye while out for a run in the rain. I wrote a book about OCD and my experiences in 2014 and in it here's how I described the change:

> My consciousness soared above my fears, as a camera draws out from a single house on a map to show the street, the town and then the surrounds and countryside. Previously, my OCD interfered with this process. No matter how much I tried to make the camera pan out, the irrational fear stayed in view, like a dirty smudge on the lens. Now the risk of HIV from all those unlikely routes shrank as I rose above to see them in their proper context. Psychologists call this moment of clarity the helicopter view. We see the landscape and all it contains in its proper scale. We regain, in all senses of the word, perspective. From 10,000 feet up, the gap between very low risk and zero risk – so visible and so important to my OCD – is hard to distinguish.

Cognitive therapies like the treatment I received are often called talking cures. But there's a lot more to them than that. Scientists now know that talking cures ease the suffering of millions of people by seeding long-term changes in connections and functions of the brain. It's a relatively new discovery, but scans of people given this type of therapy show it strengthens the wiring between parts of the brain.

And people with brains that respond with the most rewiring show the largest improvements in their symptoms.

The new connections help these people to access or tune into a part of the functioning of their brain that they previously couldn't. They improve cognitive performance. But the change – the forging of these connections in the brain – is difficult to predict. Some people respond better and faster than others and so, given the constraints on resources and the difficulty in accessing treatment for mental illness, the sad reality is that some people given cognitive behavioural therapy for all sorts of conditions don't get as much relief as they could.

To increase the success rate of this treatment, doctors and scientists look for ways to make the brain more receptive, more malleable, so the same dose of medicine and the same number of sessions of therapy can have a stronger effect. It's a new science, so the techniques are largely confined to experimental trials so far. Those techniques try to alter the way the brain makes and forms connections and rely on two main strategies: drugs and stimulation with magnets and electricity. The drugs include compounds already known to enhance cognition and brain function, such as modafinil, which makes people more alert and is prescribed for sleep disorders. The stimulation sees small electric currents wired directly into (or induced within) the brain, to artificially alter the way the brain neurons activate. (In some trials these techniques are used alone – they don't seek to improve conventional cognitive treatments but to replace them.)

4

In recent years, medical journals have filled with case studies of seemingly miraculous cures based on these new techniques: a pregnant woman freed of depression with electrical stimulation here, and a young man lifted from catatonic schizophrenia there. As word of these successes spreads, more psychiatrists, scientists and medics are turning to these cognitive enhancement techniques to try to ease the burden of the one in four people in the world who struggle with a mental disorder.

But what about the remaining three-quarters, those people who are currently healthy? If drugs and electrical stimulation can help steer and form brain connections, then couldn't everybody benefit from them? There is a strong tradition, after all, of drugs and other medical interventions being used by people not to treat a disease, but to enhance their performance. The use by athletes of medicines like steroids to build muscle – to make them stronger and swifter – is the most obvious example. As doctors and patients are now using these cognitive enhancement methods, then what is to stop everybody else doing so as well?

Do they work? Can they help us find and use parts of our brains that were previously off-limits? If so, is such brain doping fair? Should it be allowed, or even encouraged? Could it increase our attention? Our memory? Our maths and language skills? Could cognitive enhancement – in other words – increase our intelligence? And if it can, what are the implications for society? It's too soon to answer all of these questions but it's not too soon to ask them. And that's what I try to do in this book.

Intelligence is like art and pornography. We struggle to define it, but we recognize it when we see it. And we argue about it endlessly. Attitudes to intelligence divide people along myriad fault lines – scientific, cultural, political and especially educational. These disagreements steer policy and determine millions of futures. They have been used as an excuse to segregate and mutilate children, to justify unearned privilege and to support discrimination, prejudice and hatred. Yet despite all the posturing and grand claims from one side or another, the realities of intelligence science are pretty simple and uncontroversial. Yes, intelligence (or at least IQ) is partly determined by genes. Yes, IQ is a pretty good average proxy for what most people consider to be intelligence. And, yes, it's misleading and reductionist to try to condense the spectrum of individual human value and abilities into a single number such as IQ.

Human intelligence is a minefield science because the intelligence of a single human is meaningless. Intelligence is a judgement on the relative differences in ability between humans, and those differences are used to rank and judge and divide. Differences in intelligence between people – real and perceived – have long been a gateway to all sorts of unpleasantness. But these differences in intelligence also open a door to cognitive enhancement. One of these differences is the savant skill.

Savants are usually people who have low intelligence by

most standard measures but show brilliance in a single cognitive zone. They are people who typically cannot dress themselves or hold a conversation, but have staggering mental arithmetic skills or can remember every word of every book they read that week. Savant skills are commonly associated with autism, but while there is a connection the two are not the same. Savants are much rarer, for one.

How savants do it is a mystery, but one popular suggestion is they have access to some functions of the brain that remain off-limits to the rest of us. The wiring and connections of their brains are set up in such a way that they can perform these feats of mental dexterity by unlocking some extra level, perhaps as compensation for mental damage or problems elsewhere. Crucially, this theory says, there is nothing inherently different about a savant brain except for the way it is used. The rest of us could find and unlock these abilities too, if our circumstances were different.

There is some evidence for this, because not all savants are born. Some are made. For these acquired savants, new skills in maths, memory and art emerge from nowhere, sometimes quite late in life and often as a consequence of the brain experiencing some kind of trauma. Disease like dementia can seemingly release unknown artistic skill, and blows to the head can produce a near-photographic memory. Some of these changes to the brain – and the subsequent rewiring – can reasonably be said to improve these people's intelligence. They show what is possible. It's important not to get carried away; despite the way these acquired savant cases are often presented (sometimes by the acquired savant themselves) the changes do not always produce genius. Not everybody can be upgraded to an Einstein or a Mozart. Talent remains rationed. But these mental upgrades do suggest that everybody's brain might be able to work better. The question is whether that brain change can be introduced in a safe, reliable and controlled way. In the search for cognitive enhancement, few of us would be willing to bang our heads on the pavement and hope for the best.

That's where science comes in. The same research that aims to rewire the brain so more psychiatric patients can find and exploit the medical benefits of cognitive treatments could give us reliable ways to boost other mental functions too. Scientists would like this research to proceed cautiously and with proper safeguards. Fat chance. Already university students are buying black-market smart pills like modafinil and using them to help with exams. Already people are building their own brain stimulation kits and self-experimenting to try to increase their memory, attention and maths skills. Already tech companies are selling ready-built consumer versions.

This book explores this frontier of cognitive enhancement. It addresses scientific and ethical questions and issues. And it does so by investigating what it is about human intelligence – and human attempts to understand, define, measure and improve intelligence – that seems to make so many people so uncomfortable. We will look at what intelligence is and where it is found in the brain. And we will see how it can be changed. For, like it or not, change is coming. Cognitive enhancement offers the promise of assistance for those who need it. And also help for those who don't. We should investigate both.

ONE

Our Brain Revolution

There are two things you might not know about the electric chair. First, it was developed by the same man who invented the light bulb, Thomas Edison. And second, he did so not to showcase his own expertise but to attack the technology of his rival, the businessman George Westinghouse, with whom Edison was engaged in a bitter feud over the future of power.

Edison was no fan of capital punishment, but he was willing to put his personal morals to one side for money. In the late 1880s, the United States was searching for a new way to execute condemned prisoners, with hanging judged too barbaric for the emerging superpower. Thoughts turned to the new power of electricity, and its new-found ability to kill. And a decision had to be made about which of the two competing types of electric current to use.

Edison's fortune rested on direct current (the DC in AC/DC). Westinghouse was a threat because his rival alternating current (AC in the above) was easier to transmit down power lines. But there was a catch – to transmit alternating current it was geared up to high voltages, and this made it lethal. For the first time in history, people were regularly electrocuted – usually workers who were installing and maintaining the high-voltage cables.

Edison saw the opportunity to label his rival's work as dangerous. He told all who would listen how Westing-house's system was too risky, and if people didn't get the message then he showed them what alternating current could do. In a series of gruesome demonstrations, he used Westinghouse's invention to electrify a tin tray, and led stray dogs onto the metal surface to take a drink from a bowl at the other end. As the dogs yelped and dropped dead, Edison told people it could be them next. But not, he smiled sweetly, if the power used to supply their homes and businesses was the lower voltage, and inherently safer, Edison Corporation's direct current.

It was a dentist from Buffalo who suggested to Edison that electricity might serve as a capital punishment. Having watched a drunken man electrocute himself when he touched a live generator, Alfred Southwick wrote to the inventor in 1887 to ask which of the two forms of current might 'produce death with certainty in all cases'. Edison wrote back that the best execution option would be 'alternating machines, manufactured principally in this country by Mr. Geo. Westinghouse, Pittsburgh'.

Westinghouse was furious and when officials in charge of executions came calling, he refused to sell them his AC generators. His protests failed. Somehow (almost certainly with Edison's help) the officials got the equipment they wanted and in 1890, an axe murderer called William Kemmler was sentenced to be put to death in the new AC electric chair. Edison, naturally, was delighted. Kemmler, he crowed, was going to be Westinghoused.

Kemmler's execution was an oddly informal affair. He was led into a crowded prison basement and introduced to twenty-five people invited as witnesses, at least a dozen of whom were curious doctors. Then he took off his coat and sat himself in the chair. Straps were tightened, electrodes plugged in and a black cloth pulled over his face. When the warden gave the order to pull the switch, Kemmler went rigid.

After seventeen seconds of current, a witness declared him dead. Nodding, the warden started to remove the electrode from his prisoner's head when another cry went up: 'Great God! He is alive.'

Though Kemmler was unconscious, the electricity had not done its job. 'See, he breathes,' one witness cried. 'For God's sake kill him and have it over,' urged one of the journalists present, who promptly fainted. As other witnesses retched, the current was turned back on, and left on.

After Kemmler was finally dead, scientists, doctors and death penalty advocates were eager to examine his brain. Among other things, they wanted to identify the cause of death, which was important to know for the electric chair to be accepted as the latest, most humane, method of execution. But, and here is something else you might not know about the electric chair, no one has been able to work out exactly how the current killed Kemmler, or any of the 4,500 prisoners who have followed him into the chair since.

Kemmler's brain looked like it had been cooked. Its blood had solidified and seemed like charcoal. The post mortem reported: 'It was not burned to ashes but all of the fluid had been evaporated.'

In contrast, other electrocuted brains showed signs of massive internal trauma, with tissues ragged like they had been shredded by disruptive force. The massive current, scientists concluded, could make the brain literally explode from the inside; perhaps because it forced bubbles of gas to form in blood.

Electricity has unpredictable effects on the human body and on the brain in particular. Exactly what the current does in there is a mystery. This is partly why the United States (and the Philippines, its former colony) remains the only nation to have used the electric chair as a form of execution. It's why several US states have banned it and why most death row prisoners, when offered the choice, opt for the relative certainty of a lethal cocktail of drugs. And it's why, in a small flat near London's Wembley Stadium in the days before Halloween 2015, when a Ukrainian man called Andrew, with a cat and a penchant for mediaeval weaponry, straps electrodes to my head and asks me if I am ready for him to turn on the power, I swallow hard before I say yes. I don't want anything to go wrong. I *really* don't want to be Westinghoused.

The human brain packs a tangle of 86 billion different cells and, if they could be counted, the number of different ways they can combine and connect would be the highest number of anything that could be counted anywhere – not just more than the grains of sand on a beach, but greater than the grains of sand that could exist on all of the beaches anywhere. As we mentioned in the introduction, you have probably heard you use only 10 per cent of your brain. That's not true. All of your brain cells and tissues are overloaded with function. Every bit of your brain does something, and most bits do several things at once. If anything, rather than having 90 per cent spare, there is not enough of your brain to go around. But it is true you probably don't use all of your brain's potential.

This is where Andrew and his electrodes come in. Andrew is part of a growing movement that interferes with the workings of the brain to try to improve it. In basements and garages, but also in universities, military bases and hospitals, scientists and enthusiasts are using techniques to hack, boost and improve the human mind, to dig into that unused potential, make the brain work better and be all it can be. They call it neuroenhancement. We can call it increasing intelligence.

I was surprised when Andrew suggested he could neuroenhance my brain with his electricity. When I had asked to visit him at his flat, I thought we were going to talk about something called DIY electrical brain stimulation. I guessed that I hadn't made the 'do it yourself' bit clear enough. But it felt like it would have been rude for me to refuse his offer. Still, as he dampened the electrodes and placed them onto the top of my head, I wasn't sure that I wanted him to turn the machine on.

'Ready?' he asked.

'Yes,' I said, thinking, no.

'You might feel a small burning sensation.'

The furniture in Andrew's place bears the mark of someone who spends a lot of time at his keyboard. Only the chair looks truly valued – the comfortable, adjustable, expensive-looking black leather chair pulled up next to his computer.

When Andrew is not sat at his computer, he enjoys martial arts and self-defence. His flat isn't big, so when I sit down I do so next to a sharp, full-scale trident. On the wall is hung a flail, a spiky iron ball on a heavy chain and stick. They are not just ornaments. Andrew tells me he regularly uses them. I'm not sure what for and I'm not sure I want to ask.

It might seem hard to believe, but the trident and flail aren't the most striking features. Everywhere there are piled boxes of gadgets and what looks like stereo equipment, but isn't. Almost all of it is for Andrew to self-stimulate his brain, with magnets, lasers and electrical current – both direct and alternating. He does it, naturally, because he thinks it helps him. And he is certain it works. When he wants to write, concentrate or just relax, he has a brain stimulator to help him. And he's equally certain the rest of the world is going to catch up and realize the benefits soon.

As I look around the flat, his cat pads gingerly past what, I think, used to be an American football helmet on his desk. It has been converted to a brain stimulator and is loaded with electrodes and wire. I ask if the helmet is from the San Diego Chargers, but I don't think he gets the joke. He asks if I want a coffee.

With the electrodes in place, when Andrew slides the switch to on, an electric current – not much, but enough to light a small bulb – pours from the small black box in his hand, through the wire and onto the top of my head, from where it passes through the skull and penetrates a good inch or so down into the top of my brain. Shocked into action, the brain cells there become easier to fire, and so more willing to work and make connections with neighbours and colleagues. These routes and circuits between adjoining neurons become fixed into place, and, lo, that tiny fragment of my brain, that splinter of my cognition, is coaxed into working that little bit better. That's the theory, anyway. In reality, nobody really has a clue what Andrew and his DIY electrical stimulation is doing to my brain, any more than they do with the electric chair and its much, much higher currents. Still, whatever it is, he does it for twenty minutes.

After Andrew disconnects me, he asks if I feel different. I think he wants me to be convinced, as sure about the benefits of neuroenhancement as he is. Maybe, I reply. But it's hard to say. I'm relieved I think, but that's not what he means. I do feel alert and acutely aware of my surroundings. But then that could always be a straightforward caffeine buzz. When Andrew had mixed my instant coffee earlier, I'm sure he asked if I wanted it made with three or four spoonfuls.

*   *   *

Each generation has the privilege to live through a scientific revolution, and ours is neuroscience. For our parents and grandparents, the revolution was genetics; the implications and possibilities of which are still being fully explored today. For their parents and grandparents, certainly anyone who grew up in the middle of the twentieth century under the shadow of that mushroom cloud, the cutting edge of science was physics. Further back, great-great-great-great-grandparents and the rest were among the first to see the major societal impacts of chemistry, and their older relatives – if they were lucky – the benefits of medicine and anatomy. (If they were unlucky they probably helped to teach it.)

Each scientific revolution changes the world in its own way. Each presents power: over our bodies, the elements, the forces of nature and our DNA. Some of the results are good and some less so. Such is the way with revolution.

Next in line for this kind of rapid change is the brain and with it the human mind; the core – the soul if you like – of what makes us who we are. And the implications of modern neuroscience once again are extraordinary. It's our generation's turn to test nature's limits and push beyond. And depending on how it plays out, our children will inherit a different world as a result.

Prior scientific revolutions followed a consistent pattern. First, scientists explore and gather information on how we and our world work: how atoms hold together, how blood circulates, how base pairs make DNA, how a mixture of gases combine to form air, and so on. And then other scientists use that information to intervene, to harness and alter natural systems for our benefit and according to our will.

So it is with our neuroscience revolution. Since the end of the twentieth century, technology to scan the brain at work has become routine. The colourful images produced claim to show the regions responsible for all manner of human characteristics, from the neural seats of love and hate, to the brain cells that determine whether someone prefers to drink Coke or Pepsi. Until now, most neuro-scientists have been largely content to watch and map this neural activity, in all its spiralling complexity. These scientists have observed one of the oldest rules of human conduct: look but don't touch.

That is no longer true. As the neuroscience revolution takes hold, and the possibilities become clear, a new generation of scientists is not satisfied merely to watch and describe brain activity. They want to interfere, to change and improve the brain – to neuroenhance it.

The human brain is taking on its biggest challenge yet: to improve the workings of the human brain, to plot and map the trillions of possible combinations and to find a way to make them work better. To set up and connect those 86 billion neurons in a way to increase memory, reasoning, problem-solving and a constellation of other mental skills – to improve human intelligence itself.

Using science to boost intelligence might sound far-fetched, but some people in high places take the prospect very seriously indeed. In the dying days of Tony Blair's leadership, British government officials asked an expert panel to look at the possible political impact. Britain wanted to know if other countries, economic rivals, might be willing to introduce national programmes to artificially boost the intellectual 'quality' of their populations. 'We were actually seriously interested,' one of the participants recalled later, 'in whether these might be strategies that could be used by other countries which perhaps value achievement more than some countries in the West, and that would put the West at an economic disadvantage.'

State-funded scientists in China have run experiments to see if pressurized oxygen chambers – the type typically used to treat scuba divers with the bends – can improve people's mental performance. Without waiting for the results, ambitious families are booking their teenagers into these chambers the night before the pivotal Gaokao school-leaving test, the traditional route to higher education and a secure career with the state. If that sounds desperate then consider this. How well we can get our brains to work on just a handful of occasions throughout our lives is enough to steer our destiny. It's not just the obvious times – the tests and exams at school and college and interviews for jobs and promotions in work – that mark someone for success or failure. Good first impressions open doors and create opportunities, and how mental ability shows itself – from verbal acuity to simply remembering names – impresses. In a busy, overcrowded world, opportunity knocks rarely and briefly, and to have and show intelligence (or what society judges as intelligence) is one of the oldest and most reliable ways to persuade others you have what it takes to be given and to exploit these chances in life.

And the reverse is true. Our lives are haunted by the times when our brains – our intelligence – let us down. It's in the memory of humiliation over a forgotten school gym kit, and the howls of laughter at the scruffy shorts you had to wear from the lost property basket. The lasting disappointment when test scores were read out and your name came much lower down the list than you expected. The look on your dad's face when, despite his hours of help and encouragement, you failed the driving test. Again. The Friday night date with the prettiest girl in the school that leaves you tongue-tied and then red-faced the following Monday when she tells her friends how you couldn't think of anything to say on the bus.

These events have a lasting impact. Labels attached to the way we use, or don't use, our brains tend to stick. Thick, bright, quick, slow, clever, dull, smart, stupid, alert, foolish, witty, dense, canny, moronic – how quickly the dynamism of mental performance gets fixed into place. How difficult the mould is then to break. How powerful is the shift from describing people with malleable adjectives to immutable nouns. She's a genius. He's a dimwit. The indefinite article has life-long power to define us; once foolish always a fool.

This happens because the workings of the brain are considered off-limits. The brain is sealed into the skull and isolated even from the rest of our physiology. The ancients put emotion and drive and ability in the heart. We still place love and courage there. But hearts are changeable. We know how to make the heart work better. If not, hell, transplants make hearts *interchangeable*.

Science and medicine have given us God-like powers to change our bodies, to improve performance and to help us maximize our physical potential. Nobody is a weakling or a fatso any more, they just haven't tried this medicine or that workout. Progress has freed physical performance from the straitjacket of the inflexible indefinite.

Not so mental ability. The performance of a person's brain, measured by how well it can perform a series of set tasks, is still regarded as immutable. Converted into scores, numbers, percentages, grades, reflexes, responses, reactions, words and actions, this mental performance is what we think of as intelligence. And a person's intelligence, unlike physical prowess – muscle mass, lung capacity, liver function, hair growth, erectile dysfunction, stained and discoloured teeth, loose skin on the neckline, unsightly skin blotches, sagging breasts, core flexibility, body mass index, hip–waist ratio, hydrostatic buoyancy and the rest – is supposed to be static. You live life with what you've got.

This supposed fixed nature of mental ability underwrites the structure of society. Differences in perceived intelligence between individuals helped to cement in place the strata of the class system, and they remain the reason that school performance is so commonly used to rank and judge and select for potential. It's why a straight-A student will still have an advantage in a job interview several decades hence. It's why educationists and sociologists tie themselves in knots trying to distinguish whether an individual's intelligence comes from their genes or environment. And it's why people can refer to a single number as their IQ, just as they would their shoe size or height.

To alter someone's intelligence, to change the output of their brain so fundamentally that they, in effect, become a different person – certainly more so than they would after a heart transplant – seems far-fetched, even impossible. Yet we live in a world in which websites allow us to choose the shape of our nose and the length of our new, post-surgery toes. So intervening to change the way the brain gives us mental ability – probably the greatest influence over the direction our lives can take – is not just likely, it's inevitable.

For who would not like the chance to exploit that unused cognitive potential? To erase that changing-room humiliation? To think of something, anything, to say to their date, instead of sitting in embarrassed silence on the bus? To see the pride in Dad's eyes as he throws you the car keys. And who would not want to give themselves and their children the opportunity to do better in those tests and exams, to record a higher score next to their name. Well, now science says you can.

Make no mistake: our neuroscience revolution is under way and gathering momentum. Nations across the world are locked in a scientific race to explore and claim territory in the brain, each pouring billions into ways to plan and plot changes to the way all of your neurons work. This revolution is driven by demand for new ways to address the gathering dementia crisis of an ageing population, and the shocking lack of reliable treatments for mental disorders that burden at least a quarter of the global population. The question is only how far this research will spread its influence into broader society.

History shows the consequence of scientific progress cannot be bottled and constrained or forced along desired paths. It seeks unmet human demand. So, the science of anatomy and life-saving surgery has been spun off to give breast implants and nose jobs. Mastery of chemical synthesis, alongside fertilizers and targeted cancer medicines, now yields recreational drugs and new legal highs. Genetic techniques tackle inherited diseases, to spare future generations the ills of the past, but they also raise the spectre of designer babies: infants selected for gender, eye colour or height.

Given that all these types of medical research on the body have been repurposed to enhance our mood and appearance, it would be naïve in the extreme to think the same will not happen with neuroscience, where the pay-offs could be much greater. We live not just in the era of the brain. We live in a time of cosmetic neuroscience, with enhanced intelligence the ultimate prize.

People have always sought advantages over their rivals. All parents want to give their children the best start in life, or more honestly in many cases, to get other people – teachers, future employers and lovers – to notice, value and favour their child more than someone else's. But, until now, trying to improve intelligence as a way to do that has been off-limits. An education can be bought, but ability? Well, you either had it or you didn't. Yet now cognitive enhancement promises that someone who doesn't have intelligence today could have it tomorrow.

The rise of neuroenhancement challenges us to think about intelligence and ability in a new way. Around the time the UK government asked experts to investigate enhancement, the Parliamentary Office of Science and Technology produced a briefing note on the topic for British policy makers.

'Widespread use of enhancers would raise interesting questions for society,' it said. 'Currently individuals with above average cognitive performance in areas such as memory and reasoning are valued and rewarded. Making such performance readily available to all individuals could reduce the diversity of cognitive abilities in the population, and change ideas of what is perceived as normal.'

Just as with doping in sports, the benefit that cognitive enhancement techniques offer does not have to be colossal to be significant. Intelligence is relative. It's like speed in the old joke about the two wildlife cameramen filming a lion. As the hungry beast notices them and gets roaring to its feet, one of the pair slips off his jungle boots and laces up a pair of trainers.

'You'll never outrun a lion,' says his colleague.

'I don't need to. I just need to outrun you.'

The rise of neuro and cognitive enhancement raises many questions, from moral and ethical to technical and societal. For now, perhaps two issues matter more than most. First: does it really work? Second: how far can it go?

This book is a report from the front-line of the neuro-science revolution. I believe it works because I have used cognitive enhancement to increase my own intelligence. I have used it to dip into my 90 per cent unused brain potential. The evidence? I used it to cheat my way into Mensa.

TWO

Mensa Material

Mensa is Mexican slang for a stupid woman. Most people know it as the name of the international high IQ society. Mensa offers membership to people with IQ in the top 2 per cent of the population. On the mostly commonly used scale, that's an IQ of 130. Not that Mensa would say so. They prefer the more Mensa-sounding criteria that a member must prove they are on the ninety-eighth percentile. For every two members of Mensa, the organization judges there are another ninety-eight people not intelligent enough to join.

There are well over a million people in the UK with an IQ of 130 or above. The membership of Mensa UK in 2016 is about 21,000. So clearly, not everybody with a high IQ wants to join a high IQ society. That made the dozen or so people I met at a London university one Saturday morning in 2015 something of a rarity, for they did want to join. Indeed, some were desperate to do so.

We were all there to sit the Mensa entrance tests, held at one of dozens of supervised sessions the society offers up and down the country each month. The others were there to join. I was there to get my baseline IQ score, before I started a self-experiment in cognitive enhancement.

Waiting outside to be called in to begin the test, we were quiet, partly because the exam conditions were seeping through the closed door ahead of us, where a middle-aged man and woman were setting out papers on rows of separated desks. But mainly it was the large SILENCE PLEASE EXAMS IN PROGRESS signs. Peering into nearby rooms, I saw other students sitting other written tests. I assumed these were more important than ours, until in whispered conversations with my fellow would-be Mensans, I realized ours was pretty important to some of them too.

One, a nervous-looking school student, wanted to put Mensa membership on his CV when he applied to university. Another said she was accepting a challenge from her family: her father and mother and older siblings were all members and now it was her time to prove herself equally worthy.

Once we sat down, we were given two separate papers, each a series of timed sets of multiple choice questions. There were more questions than time – thirty to be answered in three or four minutes, that kind of thing. It didn't pay to hang around and ponder them for too long. But on the other hand, they got progressively more difficult, so skipping and moving on didn't seem a good idea either.

The tests of the first paper were symbols and shapes: the odd one out, next in a series, what does it look like if you rotate in this direction – the style of puzzles I had assumed we would be given. But they were hard. I got barely two-thirds of the way through the first set of questions before the blonde-haired woman in charge told us to stop writing. When she wasn't watching, I ticked A for the rest. Well, I justified to myself, I was going to use cognitive enhancement later to cheat anyway.

As the tests and the questions continued I got quicker, but I didn't feel like I got better. The dots and dashes and squares and triangles and instructions on what do with them became a language I simply didn't speak.

Then the first half was over. I couldn't read the faces of those around me to judge whether my reaction – relief and shock – was typical. I knew from our talk before at least a couple of them had tried and failed the test already. To know what was coming would have definitely helped, I thought.

The second paper swapped the symbols for words. The format was the same – several sets of timed multiple choice questions of increasing difficulty – but the focus this time was language. Some words had to be defined, others placed into context or used correctly to complete sentences and paragraphs. This was more like it. As a journalist, over a near-twenty-year career I have written, edited, proof-read or drafted, conservatively, one newspaper or magazine article a day. Say 250 a year, or 5,000 in total. A thousand words each? Five million words have passed through my brain, been sorted, queried, rejected, spell-checked, swapped, deleted, reinstated and ultimately used. And that's just on work-time.

The Mensa wordy questions weren't easy, but they were solvable. Is 'separate' the equivalent of 'unconnected' or 'unrelated'? Or 'evade' – is it the same as 'avert', 'elude' or 'escape'?* I felt like I was using a different part of my brain than in the first test. Rather than trying to eliminate the wrong answers as I had been doing with the symbol questions, which helps to explain why they took so long, I found I could more often directly pick out the solution, the right word. I even finished one of these linguistic tests early and as I put down my pen, I wondered how the girl whose family had challenged her to join Mensa was finding it. I couldn't be sure, but I thought she had sounded like she was German. I didn't catch up with her afterwards to ask, but from chatting to the others and overhearing their conversations, it seemed I was unusual. Consensus was the second test had been much harder. As we said our goodbyes, I kept my five million words to myself. The test, by the way, cost £25. And no refunds should you fail.

Until a few generations ago, the idea of a high IQ society such as Mensa would have seemed bizarre, or even more bizarre than it does now. Before the twentieth century, few

---

* Mensa is understandably reluctant for people who sit its tests to share the questions afterwards. I include these examples only because they are already in the public domain, having been published in previous newspaper articles.

people cared how intelligent they were. And they were even less bothered by how intelligent other people were, which of course is why those other people rarely cared either. School for most was a luxury when there was work to be done. Social mobility, the idea that talents and mental abilities could dictate who did what rather than the social status of family, was held back by rigid class-based rules of engagement, and when people did manage to 'better themselves' it was practical skills that usually counted.

One of the first nations to take intelligence more seriously was France. In the late nineteenth century, France was still smarting from the loss of the Alsace-Lorraine region during the Franco-Prussian war. The French government wanted it back. And it was willing to play the long game. Many in France pointed to the way Prussia had introduced compulsory primary education and had been forcing generations of its youngest children to attend school for at least a century.

France decided it too needed to create a new generation of bright, resourceful and educated soldiers. It wanted to launch a national cognitive enhancement programme. So in 1882 France followed the Prussian lead and made it compulsory for all young children to attend primary school.

Teachers in these new schools were stunned. Large numbers of their pupils appeared unable or unwilling to learn. These teachers were some of the first to wrestle with a social problem that has split the field of education ever since: how to teach a class of children of mixed ability, while not ignoring the different needs of the children at the top and bottom.

To work things out, the French created a ministerial commission to investigate and report back with recommendations. The commission was headed by a senator called Léon Bourgeois (who, despite the name, was a radical socialist) and he appointed to his panel two deadly rivals: a then-famous psychiatrist called Désiré-Magloire Bourneville, and a now-famous psychologist called Alfred Binet. Their deliberations would fire some of the first shots in a battle that continues to this day between psychologists and psychiatrists over the best way to understand the human brain and what it is capable of.

A senior figure at the Sorbonne University in Paris, Binet became fascinated by cognition and intelligence after he and his wife had two children, Madeleine in 1885 and Alice in 1887. The two girls, Binet noticed, learned at different rates and in different ways.

Madeleine as a child was cautious, thoughtful and picked up ideas quickly. Alice was more outgoing and took more chances – she would give up something she was holding before she had decided what to grab next, in a way Madeleine never did. The contrast between the girls sharpened as they grew older.

The two year gap between his children gave Binet a home-made laboratory to test mental ability and how it related to age. For instance, both Madeleine and Alice started to use the word 'I' instead of 'me' when they were three years old. His younger daughter did not make the transition any sooner, even though she had her older sister to copy from. Yet there were other skills they were equally adept at. Both could distinguish the longer of a pair of lines quickly and correctly; in fact they could do it as well as their scientist dad.

Binet thought his observations of childhood cognitive ability could solve the problems faced by teachers in the new French school system. If the children struggling to learn could be identified early, then measures could be taken to help them. Most accounts of his life paint him as an altruist, a well-meaning figure who was sympathetic to the extra needs of the disadvantaged children who struggled to learn. In fact, a little-known paper he published in 1905 on 'the problem of abnormal children' shows he was also concerned about the possible threat the less intelligent kids posed to civilized society.

If these children were excluded from school, he wrote, they would turn to crime and become a burden for the more able. 'They become parasites that consume, without any benefit to society, the work of hale and healthy men,' Binet warned. The solution, he said, was to keep them in the education system, where they could be supervised and not tempted astray by bad advice, which their inadequate intelligence would be unable to resist.

Binet had another motive too. He wanted to stop psychiatrists like his rival Bourneville being handed the responsibility – and getting the credit – for figuring out what to do with these children. The psychiatrists were most interested in severe cases, children who struggled to learn and who could be treated as a medical problem in special asylums. Binet instead saw an opportunity for psychologists – himself and his friends – to work with schools and educators on what he insisted should be viewed as a social problem.

To find the children who needed help, Binet needed a way to distinguish them from the rest. Remembering the differences between Madeleine and Alice, he designed tests of how children developed with age. A four-year-old child's performance could then be compared to what most four-year-olds could do, a five-year-old measured against the typical performance of five-year-olds, and so on. To do this, he drew up a scale of thirty different tests, which – like my Mensa tests – got progressively harder.

According to Binet's scale, most three-year-olds should be able to point out their eyes, nose and mouth, and tell a teacher their surname. By five, most were expected to copy a picture of a square and reconstruct a card cut diagonally into two pieces. At eight, they should be able to count backwards from twenty and know the date. The tests ran to age fifteen, by which stage most kids were expected to find three rhymes for a given word and to repeat a seven figure number.

The tests were not passed or failed – it was rare for a child to succeed at all the questions up to a given age group and to fail all the ones above. More commonly, their performance tailed off as they tackled the tougher questions aimed beyond their age, and this was allowed for in the way the results were totted up. A seven-year-old who answered all of the seven-year-old questions, but also half of the questions aimed at eight and nine-year-olds, was judged to have an intelligence above average. In this way, Binet invented the concept of mental age. The seven-year-old above, through a calculation to weight the answers, was said to have a mental age of eight. One who struggled with the questions for seven-year-olds but managed most for a six-year-old had a mental age of six.

It sounds crude and Binet knew it. He didn't mind. He was using it for a specific task: to help teachers identify kids who scored significantly below the average for their physical age, and offer assistance so they could catch up.

He warned against reading too much into the mental age number, which he insisted was a score not a true measure. The test was just one of several factors teachers should use, he said, alongside the child's reactions, behaviour and other characteristics.

Most importantly, Binet stressed, not every 'normal' child would pass the tests appropriate for their age. Not every six-year-old was supposed to pass all of the tests for six-year-olds and so on. In fact, the test was set up so at least a quarter of every age group would not reach the standard of their peers. (He had identified the tests as those which 65–75 per cent of children a given age could master.) The mental age he devised, by definition, was a statistical tool loaded with caveats and disclaimers. It was a snapshot of a child's performance on a given day – and that day alone – and not a firm measure of ability or future potential.

Binet died of a stroke in 1911 and was buried in the

famous Montparnasse cemetery in Paris, a final home for writers and intellectuals and the later resting place of Samuel Beckett, Susan Sontag and Jean-Paul Sartre. It was a bit less crowded in 1911, which was useful for Binet as he needed the room. The way his system was used in the decades after his death, with his caveats and cautions ignored, would see him spin in his grave for years.

Alfred Binet's work laid the foundations for IQ tests like those that Mensa use. Forget the simple lists of questions you can find online. The best modern IQ tests cost hundreds of pounds and take several hours to administer. Their questions are a closely guarded secret: only accredited psychologists and other experts can get hold of them. They probe a range of cognitive abilities, from language and mathematics to spatial awareness and short-term memory. And, more than a century on, they still rely on the basic scoring principle established by Alfred Binet from observations of his own children. An IQ score is a relative measure: it is your performance compared against typical performance of your peers. IQ, in other words, is a way to rank people, to place them in order of intelligence.

Critics of IQ tests, and there are many, like to point out it's ridiculous to try to reduce the myriad abilities and potential of a person to a single representative number. They're right, but it's not clear who they are really arguing with. It's much harder to find someone, at least someone who fully understands them, who truly believes IQ tests should be used that way.

Then there are those critics who scoff at the idea of IQ at

all. They typically crop up in the online comments sections beneath newspaper articles about bright teenagers who have tested unusually high. IQ tests do not measure intelligence, these keyboard experts insist, not real intelligence. But, as we'll see, it's hard enough to define what intelligence is, without trying to work out what it isn't as well.

These complainants are correct on one point: the only thing we can say for certain about an individual's IQ is it reflects their performance on IQ tests. And, to continue the circular logic, IQ tests, we can confidently say, are a good way to measure that person's IQ. But that misses the point. IQ isn't so much intended as a measure of individual ability, but a way to compare differences in that ability. And, on average, better performance on IQ tests does indicate higher levels of achievement in the wider world.

First, and most unsurprising given the pen-and-paper style of most IQ tests, students with higher IQ scores tend to spend more time in education and achieve better grades. Are these people only book smart, and not street smart? It seems not – the same positive association shows up in the workplace. The employees who are judged the best performers and managers by their bosses and colleagues are most likely to be those with higher IQs. This applies to all sectors, from white-collar highly skilled professional work to low-complexity blue-collar jobs. The trend is most striking in the military where recruits with higher IQs are most likely to do well in training.

Performance and pay are linked, and sure enough, those with a high IQ tend to earn more money. And they are healthier. Those bright teenagers profiled in the newspapers? As they grow up they are less likely to suffer from high blood pressure and heart disease, less likely to be obese, and less likely to have a psychiatric disorder needing hospital treatment. They will probably live longer. Some studies suggest a relatively low IQ carries the same extra risk of an early death as smoking.

There's more. High IQ is linked to creativity, musical ability, securing patents and winning artistic prizes. The higher a person's IQ, the less likely they are to hold racist and sexist beliefs. They are less likely to be religious and more likely to be interested in politics. They are less tolerant of authoritarian attitudes. They are more likely to be, as the marvellous auto-insult-generators popular on right-wing US websites might construct, bong-smoking, flag-burning, commune-dwelling, troop-slandering, tax-wasting, owl-kissing, Hollywood-humping, moonbat-crazy leftie liberals.

Still, despite the general link between IQ and what might be loosely considered as success in life – good grades, higher salaries, better health – there are some careers where high intelligence is considered a no-no. One of these – and make up your own joke here – is the US police force. In 1999, a would-be cop in Connecticut was told his performance on the force's intelligence test was too good, so he was rejected.

Police superiors were worried Robert Jordan, who had a degree in English literature, would get bored on the job and leave for a career with greater cognitive challenges, so wasting the money they spent on his training. Jordan challenged the decision in court but lost, on the grounds he was not discriminated against. The police had a right, the court said, to turn down people who scored too high – as well as too low – on their entrance exams. He worked as a prison officer instead.

It took at least a couple of weeks for my Mensa results to drop through our letterbox. The word Mensa was clearly visible through the envelope, so it took at least a couple of seconds for my wife to 'open it by mistake'. She called me at work with the bad news.

'Ha, you got in. I knew you would.'

This was terrible. How could I use cognitive enhancement to cheat my way into Mensa if I was already in?

Then a second thought struck me. I had got in; I truly was Mensa material. I felt a surge of pride and then was immediately ashamed. I told a colleague, and in doing so realized there is no way to tell people you have got into Mensa without coming across as smug and a bit odd. How do you know if someone you meet at a party is in Mensa? They will tell you.

What now? Maybe, I thought, I would have to set my cognitive enhancement sights higher. Mensa admits people it regards as being in the top 2 per cent of the population – one person in every fifty – but other, much more exclusive clubs exist. The members of these elite groups probably look down on the Mensa crowd as, well, a bit thick.

Under half of the Mensa membership, for example,

would get into the Top One Percent Society (TOPS). And fewer than one in ten of those TOPS members would make the grade at the One in a Thousand Society. Above that the names get cryptic and the spelling freestyle.

There's the Epida Society, the Milenija, the Sthiq Society and Ludomind. The Universal Genius Society takes just one person in 2,330, and the Ergo Society just one in 31,500. Members of the Mega Society, naturally, are one in a million. The Giga Society? One in a billion, which means, statistically, just seven people on the planet are qualified to join. Let's hope they know about it. If you are friends with one of them, do tell them.

At the top of the tree is the self-proclaimed Grail Society, which sets its membership criteria so high – one in 76 billion – that it currently has zero members. It's run by Paul Cooijmans, a guitarist from the Netherlands. About 2,000 people have tried and failed to join, he says. 'Be assured that no one has ever come close.'

I took a closer look at my Mensa test results. I was right. I hadn't passed the first test at all. But Mensa was also right, because, according to the organization's rules, I didn't need to. To join Mensa, applicants need pass only one of the two separate papers. And my score on the second, the language, was high enough.

On the first test, the symbols – more properly known as the Culture Fair Scale – I scored 128 out of 183, putting me on the ninety-sixth percentile: pretty good but not high enough for Mensa. On the second, the words – or the Cattell III B scale – I scored 154 out of a total 161. That, the

Mensa letter said, was bang on the ninety-eighth percent-ile. So I was in, if I wanted in. I just needed to pay my £50 subscription.

I figured, why not? Now I had my exact IQ test results, I would do my cognitive enhancement experiment as planned, then resit the Mensa entrance tests and try to improve them. I just wouldn't tell them I was already a member. The first test – the symbols and abstract reasoning – felt much more like a true test of natural brain power, so that became my goal: to improve my score on that test with the help of neuroenhancement.

But I was going to have to wait. Taking an IQ test for a second time comes with a built-in improvement, because the questions and the responses required are more familiar. It's hard to be sure how big this retest effect is, or how quickly it wears off. Some reports say it can be up to ten points and fades after three months, while others say it takes six months. To be safe, I decided to wait a year, which is how long Mensa asks people who fail to get in to wait until they try again.

That gave me plenty of time, I thought, to find a reliable way to increase my intelligence and so boost my IQ. Except, it turns out that it's not as simple as that. For starters, there isn't even a reliable way to define what intel-ligence is. Intelligence is a slippery concept that we all recognize but many struggle to pin down. And that means my goal of increasing intelligence – cognitive enhance-ment – is an equally tricky thing to identify. Is it enough to have an increase in IQ score? People have argued about this stuff for decades and I had twelve months to investigate. Let's start by turning the question around. How intelligent are you?

THREE

A Problem of Intelligence

A question follows to test your intelligence. It's a simple question about a group of ten men. First, some information:

One of the men has a moustache.

Three wear glasses.

One is bald.

The bald man does not have a moustache.

Ready?

The ten men shake hands with each other. How many handshakes are there in total?

Take your time. But don't take too long. While most people get it right eventually, some manage it faster than others. We are, after all, measuring your intelligence here.

When you have an answer, turn over.

Puzzle convention demands the answer be printed upside down at the bottom of a page towards the back of this book. To speed things up, it's here: forty-five.

Why? The first man shakes hands with nine people, the second man with eight people and so on. If your answer was fifty-five then you probably had each man shaking hands with himself. If you said ninety, and many people do, then you forgot once the bald man has shaken hands with the man with the moustache, then the man with the moustache doesn't need to shake the hand of the bald man. If you said a hundred, most likely you panicked and multiplied ten by ten.

If you said forty-five then well done you. But, how long did it take? A few seconds? Under a minute? Longer? The quicker you solved the puzzle, the more intelligent you are – at least according to standard measures.

It's impossible to boil down an intelligence test to a single question, but the puzzle above might be as close as we can get. It tests logic and reasoning, mathematical ability, spatial awareness, reaction speed, and the ability to ignore irrelevant and distracting information. (The glasses, moustache and baldness are classic red herrings.)

But solving even such a simple puzzle needs cultural awareness. You must recognize the term handshake for a start, and understand it is done between two people. And, more subtle, you must realize – not explicitly stated in the question – each pair need shake hands only once. What looks simple turns out to be more sophisticated. The same is true of intelligence.

Even the most intelligent people struggle to pin down what makes them seem so clever. Back in 1921, the editors of the *Journal of Educational Psychology* enlisted fourteen of the world's leading experts across relevant academic disciplines like psychology and philosophy, and asked them simply: what is intelligence?

The results were telling. Two of the experts refused to respond, one on the grounds the question was boring and the other because it was impossible. The rest sent back eleven different responses between them. Each of their definitions makes sense in its own way. Yet, many are very different. Some, it could be argued, oppose each other.

One expert said intelligence was the ability to use facts and truth. Another said it was skill with abstract thought. Several defined intelligence as a capacity – for quick responses perhaps, or attention and adaptability. Intelligence, one expert said, was capacity to acquire capacity.

Some definitions were simple: knowledge and knowledge possessed. Some were complex: the capacity to inhibit an instinctive adjustment, the capacity to redefine the inhibited instinctive adjustment in the light of imagined trial and error, and the capacity to realize the modified instinctive adjustment in overt behaviour to the advantage of the individual as a social animal. One supplied definition just listed traits: sensation, perception, association, memory, imagination, discrimination, judgement, and reasoning.

You might think more recent experts, with decades more research and experience to call on, would make a better job of agreeing a definition. But you would be wrong. A follow-up survey carried out by psychologists in 1986 showed opinions on the nature of intelligence still all over the place – though the range and themes of suggestions were markedly similar to the responses of the original 1921 exercise.

More recently, experts in Switzerland published in 2007 another set of definitions of intelligence, which by then had swollen to some seventy-odd different interpretations. These included the ability to adapt to an environment and profit from experience and to be aware 'however dimly' of the relevance of one's own behaviour to an objective. One researcher said intelligence means getting better over time and another defined it as the mental ability to sustain successful life.

The Swiss researchers went further than listing the definitions. They collated them, looked for the most common words and themes, and then tried to squeeze them into a single pithy phrase. I think they did a pretty good job: 'Intelligence measures an agent's ability to achieve goals in a wide range of environments.' They call it an informal definition, but here's an even more informal version: Intelligence is using what you've got to get what you want.

It's not only psychologists and philosophers who search in vain for a reliable and agreed way to describe and constrain the concept of intelligence. Computer scientists have long wanted to identify the telling features of human intelligence so they can build a machine to replicate them.

This quest to develop artificial intelligence began at a specific, invitation-only meeting of minds at Dartmouth

College in the US state of New Hampshire. For two months in the summer of 1956, ten experts in electronics, language, mathematics and other specialist disciplines thrashed out the possibilities and tried to lay the foundations for a project 'to find how to make machines use language, form abstractions and concepts, solve kinds of problems now reserved for humans, and improve themselves'.

More than sixty years later, progress on artificial intelligence has been slower than they expected. But then their meeting, and their whole premise, was based on an assumption we can now see was fundamentally flawed. Even before the meeting was convened, its organizers had promised: 'The study is to proceed on the basis of the conjecture that every aspect of learning or any other feature of intelligence can in principle be so precisely described that a machine can be made to simulate it.'

That can be done in principle perhaps, but in practice, as we've seen, it's proved almost impossible to even agree what the features of human intelligence are, never mind to precisely describe and copy them. That might help to explain why, while critics of the scary possibilities of artificial intelligence warn of self-aware machines and sentient computers, the best robots at present can barely fold a towel.

Even the origins of the term intelligence demonstrate the difficulty scientists and philosophers have had to pin it down. We derive the word from the Latin *intelligere*, which is a translation of *nous* – the ancient Greek term that Aristotle and Homer used to describe how a human mind determines what is real. In its original use, *nous* had religious and metaphysical meaning (the legacy of which remains in descriptions of God as the Divine Intelligence), and so Enlightenment writers in English, who wanted to portray a more science-based mental ability, ignored the spiritual-sounding *intelligence*. They preferred to say someone showed empirical *understanding*. The ideas of intellect and intelligence then became reserved for success in scholarly pursuits and academic questions, rather than an ability to do anything useful.

In response to this complexity, some have simply given up on the search for a reliable and useful definition of intelligence. Intelligence, philosophers have concluded, is the absence of a lack of intelligence.

(Psychologists have that one covered too. What makes an action unintelligent? People tend to describe three types of stupid behaviour. The first is confidence not supported by the necessary ability. Second is a failure to pay attention. And a hat trick of stupidity is completed by a lack of control. All of which would make another definition of intelligence someone who does not do any of those things.)

One problem with trying to define intelligence is it's difficult to do so objectively. Our judgement depends as much on our experiences, culture and values as it does on neutral facts everyone can agree on. That means a society's view of intelligence tends to reflect what it believes is important.

In Douglas Adams' *Hitchhiker's Guide to the Galaxy* series of books, both humans and dolphins consider themselves the earth's most intelligent species. Man thinks he is the brightest because he invented civilization and work and war while all the dolphins have ever done is splash around in the sea and have a fun time. The dolphins believe they are smarter for the same reason.

Across the world, intelligence depends on context. Western cultures think intelligent people can sort and categorize ideas and participate in rational debate. They value quick responses and speed of mental processing. In contrast, societies in the east – China and Japan included – think intelligence helps people fulfil social roles and identify and deal with complexity. These cultures traditionally prefer depth over speed of thought, and can view rapid solutions with suspicion.

A school pupil who stays silent during lessons? In Britain and Europe, teachers would think they lacked knowledge, while at least one tribal people in Africa see speaking less as a sign of distinction and ability. In English, synonyms for intelligent – bright, sharp, and incisive – indicate confidence, bullishness almost. Yet the Zimbabwean word for intelligent, *ngware*, also means cautious and prudent.

Who, for example, shows the greater intelligence in the following exchanges?

In the 1930s, the great Soviet psychologist Alexander Luria wanted to test the intelligence of peasants in central Asia. To do so, he tried to measure their ability to reason, to work out the answers to abstract problems. This was important, Luria thought, because it was not fair to ask questions of knowledge and education, since they had little of either.

To make his questions consistent with the culture of the Asian peasants, Luria tailored the content to reflect their experiences. So, in his tests of reasoning, Luria would first say to them: 'In the far north, where it snows, the bears are white. Nova Zemblya is in the far north, and it is always snowy there.'

Then, he asked the question, 'What colour are the bears in Nova Zemblya?'

It's a relatively simple test of ability to sift information and pull together the relevant bits, but it's also hypothetical and so tests ability to think in the abstract. And the lives of these peasant people did not demand abstract thought. With no direct relevant personal experience of a situation, they could not mentally put themselves into it.

As a result, their answers look facetious. But they simply couldn't process his request or what he wanted from them.

One said: 'How should I know? I have never been to the north.'

Another added: 'Why are you asking me? You have travelled and I have not.'

And a third replied: 'So-and-so said the bears were white. But he is always lying.'

Luria tried again, this time testing their ability to conceptualize and sort objects into groups, by asking them which of a list of objects – hammer, saw, hatchet, log – did not belong. But, again, for the peasants to group three together as tools was alien, irrelevant even.

One responded: 'They all belong. You need the saw and the hatchet to cut the wood and the hammer to hammer it.'

When Luria added somebody else previously suggested the log did not belong, the peasant replied: 'He probably has plenty of firewood already. But we do not.'

While psychologists have never agreed on what intelligence is, they are convinced it is something real; something that dictates behaviour and performance and can explain the differences they see between people. And psychologists are extremely confident that, whatever intelligence is, it can be measured.

Among the first to try was a scientist called Francis Galton who, in the great tradition of gentlemen scientists of the past, appears from a twenty-first-century viewpoint to have been both brilliant and a fool. And, in a mark of the privilege afforded to him and his kind, he seems to have enjoyed a haphazard career made up as he went along.

Pushed into medicine by his father, the young Galton was unnerved by the screams from the (pre-anaesthetics) operating table and switched to mathematics. After a nervous breakdown, he took off to Africa and embarked on what would today be called a gap year, or gap years. He shot hippos on the Nile and rode camels across the desert. He taught himself Arabic and caught a venereal disease from a prostitute.

Galton took with him on his journeys a pretend crown, a stage prop from a London theatre, which he intended to award to the 'greatest or most distant' person he met. It ended up perched on the considerable head of a local tribal king in what is now Namibia, who Galton described as the fattest man in the world. (The king sent his naked niece to

Galton's tent for the night as a thank you, but Galton, wearing a white linen suit, was so concerned the butter and red ochre she had smeared on her body would leave a stain that he sent her away.)

When Galton returned to London he wrote a book on how to survive the African bush, and then decided he fancied being a scientist. He had always loved measuring things (in Africa he once used a sextant to appraise the figure of a native woman from a distance) and – in what remains a popular science obsession – he tried to derive equations on how to brew the perfect cup of tea. He drew up the first proper weather map and the first (and hopefully last) chart of the distribution of ugly women across the nation. (Sorry Aberdeen.)

He turned his urge to measure onto people when Charles Darwin – his cousin – published the theory of evolution. The progression of tortoises and finches was all very well, but Galton was fascinated by the inheritance and selection of human characteristics. He wanted to trace the origins, not of species, but of people's mental performance.

Among his offences against modern values, Galton was undoubtedly a sexist in a sexist society. He pointed out how well-paid positions based on sensory discrimination, such as piano tuners, wine tasters and wool sorters, were all held by men. The (unquestioned) mental superiority of males, he reasoned, must show itself in better eyes, ears, noses and the rest of the senses. At the other end of the scale, the slow and sluggish responses of mentally retarded people, Galton said, were down to defective sensory functions.

'The only information that reaches us concerning outward events appears to pass through the avenue of our senses,' he wrote. 'And the more perceptible our senses are of difference, the larger the field upon which our judgement and intellect can act.'

So his way to probe intelligence was to test the senses. He designed and built wood and metal devices to test reflexes and reaction times. He got volunteers – and there were thousands of them – to punch targets, read at a distance, distinguish similar colours, pull, squeeze and breathe forcefully.

As they did so, Galton recorded their head size, height and weight and occupation, and became convinced he could find a way to link high-quality senses to those he viewed as high-quality people. (Then Prime Minister William Gladstone was one of the volunteers to visit Galton's grandly titled Anthropometric Laboratory in London.) The results, however, stubbornly refused to fit his idea. That was one problem with some of Galton's science. Facts just kept getting in the way.

Galton laid some shaky foundations for the investigation of human intelligence. But, although he never found an explanation for the differences between people, he believed he was on to something important. Others did too, and had more luck proving it.

* * *

Charles Spearman would become one of the most prominent scientists of his generation. But he started out as a
soldier. He won a medal for his service in Burma, yet later
described his fourteen years in the army as, 'the greatest
mistake of my life, [based on] the youthful delusion that
life is long'.

Spearman's true calling was psychology and studies of
intelligence. He wanted to reclaim for intelligence research
some of the scientific reputation it lost when, he said, the
Enlightenment philosophers had 'abandoned [it] to the
psychology of the streets'.

Spearman built on Galton's tests of sensory discrimination and intellectual achievement and, like Alfred Binet in
Paris, he focused on children. His approach also seems
haphazard but this was the early twentieth century and the
procedures of science and research were different then.
Today, researchers are typically funded by the government
and accountable to various levels of officialdom. Each project leaves a paper trail – grant applications, reviews, ethics
approval and consent forms filled in by any volunteers. It
was easier for Spearman. When he wanted some children
to study, he simply walked a hundred yards around the
corner to his local village school and asked to borrow
some.

Over several months, two dozen of the school's oldest
children visited Spearman's house, in the village of Apple-
ton near Oxford, where he spent fifteen minutes testing
their eyesight, hearing and how well they could distinguish the weights of two objects. The eyesight test was

particularly problematic to set up. Spearman wanted the children to say which of two cards had a darker shade, and he tried to make the test as fair as possible, which meant he had to place the cards each time by hand an equal distance and angle from the centre of an evenly lit window.

Next, Spearman gathered another three dozen of the school's next-oldest kids in one of the school classrooms and played to them a series of pairs of musical notes on a home-made device called a monochord. For each pair of notes, each child simply had to write down a 1 or 2 to indicate whether the first or second of the two sounds was the higher.

There they were. A full room of six-to-ten-year-olds bunched together and asked to sit quietly while a balding former soldier used a home-made instrument to play the most boring music in England. What could go wrong? As Spearman noted dryly in his subsequent report (still considered a classic in psychology): 'Energetic measures were found necessary to prevent cribbing.' With help from the school's headmaster, several teachers, and a 'small prize offered to stimulate attention', Spearman somehow got the class to sit still long enough to get the results he needed.

Next, Spearman headed across the village and up the social ladder, to a second school where he expected to find kids of a higher calibre. It was a 'preparatory school of the highest class' that trained boys for the top private school at Harrow. Arranged at short notice, the session was a disaster. With no time to test them individually, Spearman watched aghast as the boys passed around his precious weights and peered at his coloured cards under all sorts of different light. He was left on his own to supervise the class, couldn't keep them from comparing notes, and struggled to get them to take the task seriously.

He had more luck when he returned with his mono-chord. This time, several of the masters stayed to keep order. The 'social standing and general culture' of the place, Spearman noted approvingly, was now 'the opposite extreme to that in the village school'.

Spearman took his sensory measurements, but he still needed a way to check them against how intelligent each kid was. He had their exam results, which was a good start, but he also wanted to include whether teachers thought each was 'bright' or 'average' or 'dull' and to rank them accordingly. Teachers told him it was impossible. No bother, Spearman replied: simply tell me which of your kids is the brightest. They were happy to do so, and after they did, he asked which of them was the next brightest, and so on, until he had what he wanted. Finally, he got the two oldest children to independently judge their friends on 'sharpness and common sense out of school'.

Francis Galton's anthropometric laboratory had tried and failed to prove the link between intelligence and performance, but Spearman succeeded because he went a step further. He devised a statistical method to prove with maths what seemed obvious at first glance: the same names tended to group at the top of the different lists. A pupil who was good at classics was also likely to be good at French. And he found the more 'thinking' was involved in tests, the closer this bundling became. The same grouping occurred at the lower end of the scale too. Children who struggled at music were also likely to score badly in English. What looked like different measures, Spearman realized, were all testing the same thing.

Whenever thinking was required, he concluded, the pupils drew on something. The brighter pupils had more of it, which could explain their superior cognitive performance. Today we might say those kids had more of the X-factor. Spearman said this capacity was 'general intelligence'. He later relabelled it the 'general factor' and then, simply, 'g'.

Remembering his time in the army, Spearman was convinced his discovery would be useful. He said it could be used to make some important judgements about a person's broader abilities from their performance at school. When he published his results in 1904 he concluded:

> Instead of continuing ineffectively to protest that high marks in Greek syntax are no test as to the capacity of men to command troops or to administer provinces, we shall at last actually determine the precise accuracy of the various means of testing General Intelligence, and then we shall in an equally positive objective manner ascertain the exact relative importance of this General Intelligence as compared with the other characteristics desirable for the particular post which the candidate is to assume.

Spearman's observation sounds routine but it is anything but. In fact, it is one of the most controversial and disputed scientific discoveries of human ability ever claimed. At a stroke, it undermines many popular beliefs about intelligence, ability and education, and emphasizes the ruthless disinterest of Mother Nature in making life fair for her children. It has been used to defend all manner of unpleasant and unscientific and illegal forms of discrimination and to underwrite hate and prejudice and snobbery. It has been used as a justification to kill people, to forcibly mutilate them, to steal them away from their loved ones, imprison them and to make them defenceless exhibits for public mockery. More than a century on, Spearman's findings are still used to put people down and to keep them in their place.

Properly named, Spearman's big discovery is known as the positive manifold. You know it better as the kid at school who was good at everything. Maths, English, French, history – whatever it was they were top of the class. Music, art and even sport – they probably excelled there too.

Under the indifferent dominance of the positive manifold, mental excellence is both rationed and unevenly distributed. And so is failure. In fact, the connection between scores is even stronger towards the lower end of the scale. The same people do badly at everything, just as the usual suspects are the ones doing the best and getting the highest grades.

Spearman's g is probably the most important scientific theory you have never heard of. One reason it's not more widely known is because it's purely theoretical. Someone's g – their general intelligence capacity – cannot be measured, or at least psychologists and neuroscientists have found no way to measure it directly. IQ scores are the closest proxy, and that's one reason why IQ is hated by so many; all the social and political baggage of the idea of a fixed general intelligence gets piled onto the back of IQ.

Among that heavy load is the tacit assumption that intelligence is valuable, and those with more of the trait are better somehow. That it's good to be intelligent and human progress should head in that direction. For is there a bigger crime than talent wasted and opportunities not realized?

Few people will admit to this bias, even fewer to allowing it to steer their behaviour and attitudes, but it exists. We see it even in our relationships with animals. One reason people argue against the eating of dogs is they are intelligent creatures – more so than the dumb sheep and cows they happily cook. The same applies to the octopus. Plenty of campaigners argue an animal bright enough to recognize itself in the mirror has no place in paella.

Importantly, the nature of intelligence is about more than philosophical definitions and academic chin-scratching. People's attitudes and beliefs on how intelligence works, and how it might be increased, have a real impact on how they view the abilities and potential of themselves and others. Most fundamentally, a belief that intelligence is a fixed quantity and can't be increased is often enough to make a child perform poorly at school, and influence how well they set themselves up for life.

In a typical group of ten school children, four believe intelligence is fixed. Whatever they do, they think they cannot change their mental abilities. Another four think the opposite – their intelligence can be improved, and the best way is to work hard. (The other two kids of the ten don't pick either option.) It doesn't matter to their grades which group is correct – whether intelligence is actually fixed or not – their belief alone steers their attitudes, effort and performance.

It is better to believe intelligence can be increased. Those children who believe the opposite, that intelligence is fixed (called the entity theory), are more anxious about how much intelligence they have, and it not being enough for them to succeed. These children refuse opportunities to learn if they carry a risk of doing poorly. They conceal or lie about their weaknesses, rather than identify and improve them. And they indulge in what psychologists call self-handicapping – procrastination and watching television the night before a test instead of studying. This gives them a ready-made excuse if they score badly.

There's more. This group believes ability alone should be sufficient to succeed. So they think effort and persistence indicate low intelligence and don't bother with either. When things get difficult, this group gives up, cheats, loses self-esteem and ultimately does worse. That's because they believe they are bumping their heads against an intellectual ceiling.

In contrast, children who believe intelligence can be improved (called the incremental theory), have a healthier attitude to their studies. They value effort as much as performance, and bounce back from failures with renewed determination. For them, the sky is the limit.

There seems no strong difference between the actual intellectual abilities of the two groups, which makes the lack of effort in those who think intelligence is fixed all the more galling. They are wasting their talents, while less able children with a more positive attitude can thrive.

For obvious reasons, experts want to understand why these different groups of kids believe what they do. One explanation might be subtle differences in phrases used to praise them when they were younger. A young child recognized for ability – 'what a great picture you've drawn, aren't you clever' – could start to attribute accomplishment to fixed traits. A child praised for process – 'what a great picture, you've put loads of work into that, haven't you?' – could take the opposite and more fruitful approach, and believe success came from effort and practice. It could matter whether praise is generic or specific. 'You did a good job on that drawing' is more likely to make a child think intelligence is incremental, and so help them, than the equally well-meaning, 'oh look, you can draw well'.

It's not just the children in schools who have these different beliefs about the nature of intelligence, teachers do too. If you struggled at maths at school, even in a single arithmetic test, and a teacher tried to comfort you by telling you not to worry because not everybody can be good at maths, then, unfortunately, that teacher probably had an entity theory of your intelligence and judged it as insufficient for maths. They probably haven't given it a second thought since, but it almost certainly changed the way they taught you, and who knows what else.

FOUR

Treating and Cheating

I became interested in neuroenhancement because it's being explored and talked about as an experimental treatment for mental illness. I was lucky with my own mental problem and got some expert help, but I have met many people with OCD and other disorders who have been less fortunate. It's not surprising that some of them are reaching for what might appear desperate measures.

Treatment for OCD and other mental illnesses is better than it used to be, but progress has stalled in recent years. Most tend to take a two-pronged approach and combine drugs, which alter brain chemistry, with psychological therapy, which encourages people to confront and challenge attitudes, thoughts and behaviours. Delivered properly and together they mostly work with most people most of the time. But there is plenty of room for improvement.

Particularly puzzling is the uneven response of patients to the psychological treatment, the so-called talking cure. Usually a type of cognitive-behavioural therapy, this tends to be offered as a series of sessions, perhaps a few hours a week for three months, typically in groups. It aims to make people aware of their own thought patterns, the bad cognitive habits, and introduce drills and behavioural responses to help overwrite them.

What puzzles and frustrates psychiatrists is that the benefits of cognitive behavioural therapy for OCD, depression and other mental disorders, come in fits and starts. Moments of clarity, epiphanies, 'eureka moments' – call them what you will – these spikes in improvement arrive at different speeds in different patients and at different times. Sometimes the benefit is fleeting and must be coaxed back, for others the change is more solid.

One reason that people might experience no benefit from cognitive therapy, and there are significant numbers of them, is that their brains are simply stuck. They could be slower to respond, the lock on the box could be tighter. And, in that case, perhaps a dose of a cognitive enhancing drug or a tickle of electricity could nudge them along, help their minds to unlock the change in state they are so desperate for.

Electricity applied to the brain of psychiatric patients conjures images of shock therapy; this ECT delivered as a quasi-punishment was made infamous by its depiction in *One Flew over the Cuckoo's Nest*. Originally delivered without anaesthetic, the massive doses of electricity were intended to induce seizures, unpredictable and uncontrollable storms of electrical activity in the brain. The muscles on the other end of the connecting nerves, baffled by the blizzard of messages from the brain, would convulse and the unfortunate patient would thrash about, often breaking arms and legs. Why induce seizures to treat depression or psychosis? The same reason that the most common advice from computer experts is to turn a crippled machine off and on again. It's the reset button, the control-alt-delete of the mind.

ECT is still out there and is used, with some reported success, to treat depression. But it's extreme and unpopular. The electric currents psychiatrists experiment with today are milder, and aim at a different target. Rather than simply reset the whole of the brain, psychiatrists now think electricity could help fix individual bugs. Applied directly to the scalp, the activity of the brain regions immediately below, they believe, can be turned up and down, even on and off. If – and it's a fairly big if – a mental disorder can be traced to unusually high or low activity in an accessible part of the brain, the theory says, then this new form of electrical control should help to relieve symptoms. It might help to unlock activity in the brain, to release some latent potential. It might help more people discover relief.

Mild electrical current passed through the skull into the brain in this way has been used to try to treat people with numerous mental problems in recent years. The results are encouraging, if hardly conclusive. There are plenty of reports of near-miracle recoveries. In at least one case, the experimental treatment has been given to a pregnant woman to avoid the possible complications of psychoactive medicines like anti-depressants. Such case studies are interesting, but subject to a unique type of academic bias: reports of

65

dramatic recoveries get written up and published, while patients who have the same treatment but don't improve are quietly forgotten. If doctors and surgeons get to bury their mistakes, psychiatrists and clinical psychologists can simply hide their failures in a desk drawer.

Controlled studies and trials are rarer, but are starting to happen. A review of studies published in 2016 found good evidence that electrical brain stimulation could help reduce the symptoms of depression and schizophrenia, and promising early signs with eating disorders, anxiety, obsessions and compulsions.

Published in the *Journal of Psychiatric Research*, the article concluded that repeated doses of the electrical therapy can, 'ameliorate symptoms of several major psychiatric disorders'. But it warns, 'The field is still in its infancy and several methodological and ethical issues must be addressed before clinical efficacy can truly be determined.'

Not everybody is heeding the warnings. Under the conventional medical model a new treatment – an experimental drug, say – is checked in clinical trials, usually with hundreds of people, to make sure that it's both safe and effective. Until it's approved, the drug is off-limits to the people it is intended to help. That's not the case with electrical brain stimulation. It can be done with one battery – typically one of the chunky ones from a smoke alarm – some wires and a bit of knowledge or instruction. As word spreads, and scientific and medical journals fill up with positive case studies, a growing number of people with OCD, depression, bi-polar disorder – desperate people who might have struggled to get good conventional treatment or for whom it didn't work – are finding ways to experiment with electrical stimulation themselves, to manipulate the workings of their own disordered brains. Some equally desperate parents are doing it to their children to try to lift the social veil of their autism.

It's not just mental disorders. Other malfunctions and misfiring circuits in the brain can be targeted with electrical stimulation too. In 2013, pupils at a special needs school in London had their brains gently massaged with electricity to see if it could help overcome their learning difficulties. Half a dozen eight-to-ten-year-olds who struggled badly with maths were given nine sessions of twenty minutes' electrical stimulation at Fairley House School, with a special cap fitted to their heads. Compared to a parallel group who wore similar caps, but with the current switched off, the brain stimulation helped them score significantly better at general maths tests.

As these studies are reported and discussed, so the use of neuroenhancement leaks beyond medical applications, and into broader society. One thing these new neuroscience techniques have in common – unlike, say, cosmetic surgery – is that they do not try to introduce anything new. They seek to release or switch on some brain potential that is already in there. That means the growing number of people who turn to these techniques are using them to find and release some hidden power within. And as they do so, they follow in the footsteps of many, many people who have gone before.

The British cyclist Tom Simpson wanted to find and release some hidden power. He wanted to find the ability to win the world's most famous bike race, the Tour de France. So, one summer's morning in 1967, he took some amphetamine pills, washed them down with brandy and headed up one of the most difficult climbs of the cycling world: Mont Ventoux in Provence. He never made it to the top.

A couple of months after I got the results from Mensa, I cycled past the memorial erected to mark the death of Tom Simpson on Mont Ventoux. Actually I passed it twice – once very slowly on the way up and once very quickly on the way down. It's at the side of the road a few hundred metres from the summit. A brutal oblong affair, it's decorated not with the usual flowers and handwritten messages but with two shelves of cyclists' colourful water bottles; the monument is entirely fitting for the surroundings. Part of the reason cyclists fear Mont Ventoux is its upper slopes are stripped of vegetation and exposed to frequent gale-force winds. The broken limestone landscape looks like snow from a distance and the surface of the moon close up. A giant red-and-white tower sits on the top and mocks as you will it to grow larger and nearer with every pedal turn.

In 1967 the amphetamines in Tom Simpson's body stopped him sweating. As he pushed towards the top, he simply overheated. Delirious, he swerved from side to side. When he fell off, he famously asked the spectators to put him back on his bike. As he slipped from the saddle for the final time he was silent.

Simpson's death caused much soul-searching among professional cyclists, many of whom at the time were taking the same kinds of drugs and stimulants. Together with new rules on what kind of help was allowed – and what for the first time was to be banned – it pushed riders and other athletes towards a new form of doping, based on science and new medicines.

We have long seen this kind of mission creep from treatment to enhancement with medicines designed and developed to fix the sick, but which were then exploited to boost the physical performance of the elite. Athletes doping in endurance sports are now the most visible example. Anabolic steroids abused to build muscle were first developed to stimulate growth and appetite in people with wasting conditions including cancer. Synthetic erythropoietin (EPO) – the banned substance of choice for cyclists in recent years – is used to treat the anaemia that often emerges as a complication of kidney and bowel disease.

Despite rules and regulations designed to stop the wider take-up of medicines by healthy people to enhance, it has proved unstoppable. And the same applies to neuroenhancement techniques being investigated and developed to treat mental illness.

Like many fringe and pioneering technologies in the past that subsequently became mainstream, at the moment the subject of neuroenhancement is most thoroughly discussed as science fiction. The best example is probably the 1966 book *Flowers for Algernon* by Daniel Keys. It's written as a series of entries in the diary of Charlie Gordon, a mentally disabled floor cleaner turned into a genius by an experimental treatment. As his cognitive abilities increase, he takes over the research project that transformed him, and calculates with dismay his new-found brain power will soon start to fade. (Algernon is a mouse who was given the treatment first and shows a vividly similar rise and collapse in brain power that is ultimately fatal.) As Charlie's self-awareness grows with his intelligence, he becomes angry and then ashamed to realize that, before the treatment, the people he thought were his friends had in fact ridiculed him.

'How strange it is,' Charlie writes, 'that people of honest feelings and sensibility, who would not take advantage of a man born without arms or legs or eyes – how such people think nothing of abusing a man born with low intelligence.' The story was turned into a 1968 film *Charly*, which won its star Cliff Robertson a best actor Oscar.

A less critically acclaimed yet still commercially successful film, *Limitless*, was also based on a book; *The Dark Fields*, published by Alan Glynn in 2001. It follows the upturn in fortunes of Eddie Spinola after he starts taking an experimental drug to increase his intellectual, creative and learning powers. After finishing his book in a couple of days, he embarks on a lucrative new career in finance. Eddie's transformation is driven by a new ability to see patterns, often in large amounts of information.

'My abiding impression of the period is how right it felt to be so busy all the time,' Eddie says. 'I wasn't idle for a second. I read new biographies of Stalin, Henry James and

Irving Thalberg. I learned Japanese from a series of books and cassette tapes. I played chess online and did endless cryptic puzzles. I phoned into a local radio station one day to take part in a quiz, and won a year's supply of hair products. I spent hours on the internet and learned how to do various things – without, of course, actually having to do any of them. I learned how to arrange flowers, for example, cook risotto, keep bees, dismantle a car engine.'

A trilogy of decent fiction about cognitive enhancement is completed by *Understand*, a 1991 short story by Ted Chiang. It features another experimental drug, this time given to a man called Leon who suffers brain damage when he nearly drowns. The drug is intended to restore lost function, but ends up increasing his intelligence massively. Leon explains:

> As my mind develops, so does my control over my body. It is a misconception to think that during evolution humans sacrificed physical skill in exchange for intelligence: wielding one's body is a mental activity. While my strength hasn't increased, my coordination is well above average. I'm even becoming ambidextrous. Moreover, my powers of concentration make biofeedback techniques very effective. After comparatively little practice, I am able to raise or lower my heart rate and blood pressure.

The three tales sound similar, but an important difference between them highlights a crucial point in the discussion of cognitive enhancement. Charlie is mentally defective and

71

his quality of life suffers as a result. Society, many ethicists would argue, has a duty to intervene if it can. The same seems true for Leon, who loses quality of life and presumably will want it restored. But Eddie, the hero of *Limitless*, is already a relatively high achiever. Boosting his intelligence helps his bank balance more than his basic human rights. To help Charlie and Leon with cognitive enhancement is treating. But is using it to help Eddie cheating?

Bioethicists have pondered the distinction between treatment and enhancement for years with physical traits. When can a medicine be given to a healthy person, to improve them beyond natural limits? It's not always easy to draw the line between the two. A common example of the dilemma describes the possible use of human growth hormone therapy for two boys of below-average height. One of the boys is short because he has a brain tumour which leads to a hormone deficiency. The second boy is short because he has short parents.

The conventional ethical model would give the growth hormone to the first boy only, because it would be labelled therapy. Giving it to the second boy would be classed as enhancement, and so not permitted. Sounds fair? Not for boy two. Various studies show being short can reduce the quality of life of men. They tend to suffer discrimination from women and employers. And what is therapy for if not to improve the quality of life?

Before Viagra was discovered to have its famous effect, no diagnosis of 'erectile dysfunction' featured in a doctor's dictionary. If a seventy-year-old man was not as vigorous as he once was, it was a lifestyle and not a medical issue and fixing it was a bonus, an enhancement, not a treatment. The drug companies weren't even looking for this solution – Viagra was meant to treat angina and hypertension. It performed poorly in this respect but an accidental side effect has made them billions.

Whatever definitions one attempts to impose on treatment and enhancement to keep them separate, logic has a nasty habit of pushing them back together again. If we regard treatment as a reversion to a 'normal' or 'average' state that would rule out heart transplants and the widespread use of statins to push the cholesterol levels in the blood of middle-aged men far below the levels possible otherwise.

This is more than just a semantic or philosophical issue. The difference between a therapy and enhancement determines real-world issues like price and access. With finite resources, the standard position is to prioritize therapy because it rights a wrong. But like 'normal', what is considered as 'wrong' constantly shifts as technology and expectations rise.

The ground becomes even less solid when the human improvements to be introduced – by both therapy and enhancement – are cognitive as well as physical, because 'normal' is much harder to define and because the likely benefits could have more day-to-day impact. As politicians are constantly telling us, we live in a knowledge economy. Knowledge is power. And a little knowledge remains a dangerous thing, especially if a political, military or economic rival has a little more. Or if they are just a little quicker on the buzzer.

In autumn 2012, I got an email out of the blue inviting me to appear on a special Christmas series of the television quiz show *University Challenge* for graduates. I am still not sure how they chose me, there seemed to be a suspiciously high number of journalists, but I was careful to reply and say yes before they realized their mistake and changed their mind.

University Challenge is famous partly because the questions are often so baffling even hearing the answer is no help to answering them, and partly because the teams are presented on screen one on top of the other. (It turns out, I learned on the day, that for one ill-fated series the producers did actually build the set like that.)

I learned something else from that day: the face I pull at the frustration of knowing an answer but not knowing it quickly enough, or of not being able to drag it from my memory at all.

I knew the phrase, *A dog is for life and not just for Christmas*, which was an answer to one of the questions. I had seen it plastered across enough car stickers in my childhood for the words to be lodged in my brain somewhere. I knew Georgia is the US state just north of Florida. Yet in both cases, when the moment arrived, I couldn't give Jeremy Paxman the answer.*

---

*Our team won, but not with a high enough score to proceed to the next round.

Would it be cheating if a little burst of electricity to my ageing brain, or the improved recall of a smart drug, had helped me remember? Advances in neuroscience mean this isn't a purely theoretical question. Scientists in New York have shown that electrical brain stimulation of a region called the anterior temporal lobe can improve the scores of students in general knowledge tests, presumably because it helped them to recall the answers to questions such as: what is the largest organ of the human body? (Answer – the skin.)

It is really 'enhancement' to help students to recall stuff they already know? If it is, then is this type of cognitive manipulation any different from the effects of a cup of strong coffee? Or the Pro Plus caffeine tablets we lived on as students a few years ago? Or the effects of a balanced and nutritious diet, no booze and a good night's sleep?

It quickly drifts from a scientific to a philosophical question, and one that returns to the debate about how to define intelligence. Is cognitive ability what we know, or what we do? Is intelligence storing information or using it? Certainly, tests of intelligence take a utilitarian approach and measure actions. Academic examinations are more aimed at probing knowledge. The difference between the two is often not a question of varying intelligence but different personalities, or simple biochemistry. The use of knowledge, for example, can be tempered by nerves and shyness. Confidence can help people to express what they know. Drugs to calm anxiety and so help worried people show their ability in exams – or television quiz shows – are those cognitive enhancers? If so, is that cheating because it is unfair on the others? If it is, then what about the unfairness of scheduling the same exam in the morning rather than the afternoon, which will inevitably put some people at an advantage and others at a disadvantage, depending on their response to the daily yin and yang of circadian rhythms? Or should we assume better control over physiology is simply another sign and benefit of higher intelligence? It does, after all, help organisms to use what they've got to get what they want.

Opinions on what to do with cognitive enhancers, whether to control, allow, regulate, ban, recommend or even just research them, are all over the place at the moment. At one extreme there are people who call themselves transhumanists, who argue we have a right and even a duty to improve ourselves, and so society, as far as possible. The model for their neuroscience revolution is Leon Trotsky's philosophy of constant upheaval.

More cautious are those who insist cognitive enhancers are risky and morally dubious. This camp includes administrators at Duke University in North Carolina who have banned the unauthorized use of prescription medicines like modafinil by students as cheating. Originally developed to induce wakefulness in those with narcolepsy and other sleep disorders, in healthy people it can improve focus, reaction times and fatigue levels, making it a student's go-to study aid. If we ban these drugs then how long before students are required to wee in a bottle and be drug tested before every exam? And is it also cheating if they use them to study harder and longer in the weeks before but go into the exam 'clean'?

If modafinil and other smart drugs can help students to focus so they can access stuff they have learned, the conventional argument says this is cheating, because other students haven't been given the same help. But don't smart drugs just help us achieve our potential, something which, after all, is one of the most commonly stated goals of education?

In fact, some evidence says modafinil could do what many people involved in education say they want: to give kids a fair and even start in life. The weaker someone's cognitive performance to start with, the more modafinil seems to help them. Now *there* is an interesting ethical question. Smart drugs only seem fair to use if everybody has the opportunity to do so. But what if not everybody benefits to the same degree, and specifically, what if those with lower intelligence are brought closer to the rest? That sounds good at first, but it's not hard to envisage some people unhappy about the shift – not least those who can currently seek the social and monetary advantages offered by an improvement in tests and exams by paying for private education.

Who should get to use cognitive enhancement? All who wish? Then what about the peer and competitive pressure placed upon classmates and work colleagues who would rather not, but know everybody else is? Or the more explicit pressure, from parents and bosses? Should we expect – and help – people who hold the lives of others in their hands to be at mental full speed the whole time? Pilots and surgeons make more mistakes when they feel tired. And judges have been shown to make decisions differently – they tend to grant more prisoners parole first thing in the morning and straight after meal breaks. For those looking to tip the scales in their favour, justice really is what the judge ate for breakfast. Should society not demand more equal treatment, and if cognitive enhancement can help to achieve it, then shouldn't we do it? Don't we have an obligation to do so?

But, on the other hand, if we expect people to pass professional tests and exams – to fly a plane or know their way around a burst appendix in a hurry – and they choose to meet those standards with some artificial help, then what happens if they decide not to take those same smart pills for a while? Does that invalidate their accreditation and remove their mandate, and if so, how is it different to a doctor who wakes up groggy after a party but decides to go to work anyway?

That's a lot of questions and not many answers. All we can say with any confidence is that cognitive enhancement is not going to go away. Smart drugs like modafinil, already available to anyone with an internet connection and a PayPal account, are just the beginning. In the background, scientists and drug companies are working away on improved, more effective cognitive enhancing drugs. And so the ethical questions above, as well as more fundamental scientific issues such as long-term safety and the reliability and magnitude of the effects, are essential to explore.

So, to explore them myself, I found a website that offered to sell me some black-market modafinil. I gave them my credit card details and ordered some.

FIVE

Pills and Skills

Suicide bombing is now a disturbingly common tactic for extremists, which makes it hard to comprehend the scale of shock, terror and confusion among the first US sailors to face the kamikaze pilots of the Japanese in the Second World War. An enemy who was willing to die to kill you, a pilot who did not fly his plane to attack with a speed and approach that allowed him to pull up and live to fight another day, was a new weapon and one almost impossible to stop.

What made the kamikaze pilots willing to die? Much has been made of their devotion to the divine Emperor, and the cultural shame of surrender to the Japanese soldier. With the ring of superior US forces closing on the home islands, the kamikaze squadrons were a final fling, a heroic sacrifice made by brave idealists who hated the enemy more than they loved their own lives. They were noble patriots who served a greater power. Less known is they were dosed up on methamphetamine. Kamikaze pilots sometimes had to fly for hours to reach the target and needed to be kept mentally alert, and the long-lasting euphoria of the drug high, their commanders assumed, would make them less likely to change their mind.

In the 1940s, Japan was heavily into methamphetamine (meth). A Japanese chemist called Ogata Akira first synthesized the drug in 1919 as a psychiatric medicine also given to people with lethargy and depression. When war came, the Japanese military government thought it might help keep its soldiers and workers mentally alert. It ordered a massive and rapid increase in production of the drug and made it available in a convenient tablet called Hiropon. May not cause drowsiness. Use of heavy machinery was actively encouraged, as the government advertised its not-so-secret weapon.

'For night work and other times demanding mental alertness. For overexertion. The most powerful new amphetamine on the market! – Hiropon tablets.'

Sanctioned use of meth as a cognitive enhancer in Japan continued after the war, when pharmaceutical companies promoted the tablets to tired workers, war veterans and those struggling to cope with the social change demanded by the horrors of Hiroshima and Nagasaki and Japan's 1945 surrender. But as reports of addiction and crime linked to the drug grew, official attitudes hardened. In 1951, meth was designated a medicine and most casual use was banned.

One group proved more reluctant than most to give up their easy route to mental alertness. High school students cramming for university entrance tests and university students preparing for their own exams continued to buy the drug on the black market. The situation became so serious that in 1954 the vice-minister of education pleaded with the heads of all universities and schools to do more to stamp out what the Japanese government now called drug abuse. These students were some of the first to use what we now call smart drugs. And, sixty years on, their use is far from stamped out. If anything, smart drugs have never been so popular.

In the autumn of 2014, UK officials raided a lock-up garage in the Midlands brewery town of Burton and came away with what they described as their biggest single seizure of smart drugs. More than 20,000 pills were found and over a dozen different types of drug. Britain was put on alert after a tip-off from Norwegian customs, which had already intercepted and confiscated several packages.

Announcing the UK haul, Alastair Jeffrey, head of enforcement at the Medicines and Healthcare Products Regulatory Agency, told journalists: 'This is a recent and very worrying trend. The idea that people are willing to put their overall health at risk in order to attempt to get an intellectual edge over others is deeply troubling.' And highly lucrative: the MHRA said the stash had a sale value of about £200,000, bought cheap from drug manufacturers overseas and marketed to students.

Despite the renewed attention from the authorities, smart drugs have continued to arrive in Britain. I know this because some of them arrived at my house. A few months after the Burton raid, they landed in an anonymous brown envelope on my doormat. The pills I had bought online were modafinil. Those who sell the pills promise it will deliver massively increased cognitive abilities.

Like other cognitive enhancement techniques, modafinil was first introduced as a medical treatment. The drug emerged in France in the 1970s as part of research to develop safer alternatives to stimulants like amphetamines, which were used to treat sleep disorders including narcolepsy but carried side effects. As modafinil was increasingly prescribed, doctors started to wonder if the wakefulness and alertness the drug promoted could help tackle fatigue and related symptoms of other conditions. They began to give it out for off-label – not officially sanctioned – use in multiple sclerosis, myotonic dystrophy and more. And as well as increasing wakefulness, scientists started to look at improvements in cognitive function.

In recent years, the off-label use of modafinil has exploded. In medicine, it is being investigated as a replacement for amphetamine-like stimulants in a range of other problems, including treatment-resistant depression and attention deficit hyperactivity disorder (ADHD). Outside medicine, the armed forces of various nations have given it to their infantry and air crews. It's believed to be so prevalent and potent that in 2015 the World Bridge Federation started to test players at international tournaments for the drug, which it considers a banned stimulant. So did organizers of the ESL One Cologne professional video games tournament. And then there are the students. It's hard to find reliable numbers but some surveys suggest as many as a quarter of UK undergraduates have taken modafinil or similar to help their work. A fifth of surgeons say they have taken it, and a similar number of professional scientists.

Availability and legal status of smart drugs differs across the world. In Columbia, modafinil is available over the counter in pharmacies. In Russia possession is illegal. In Britain it's a prescription-only drug, so legal to possess, but illegal to sell and supply. Is it OK to receive a couple of dozen sent through the post from India? Let's call that a grey area.

Online sales of modafinil and other smart drugs are booming, but buyers beware – there are more than legal risks to consider. As the MHRA says: 'A huge number of medicines bought online are counterfeit, substandard or adulterated. There is no guarantee the product you are receiving isn't laced with any number of other dangerous substances.'

The counterfeit medicine market is huge and regularly kills people, from heart patients in Pakistan to those taking tainted steroids and blood thinners in the US. More likely, the drugs simply don't work. Counterfeit modafinil is often little more than caffeine tablets. So I wanted to verify my online smart drugs were genuine. But, as it turns out, it's a lot easier and cheaper to buy medicines online than it is to check if they are legitimate.

The brown envelope they were posted in was stamped with the name and address of a company in the Fort area of Mumbai. The same name and address was on the white sticker on the back, which was the customs declaration. Under 'Quantity and detailed description of contents (e.g.

two men's cotton shirts)' was printed 'Sample Harmless Medicine'. A tick box indicated the medicine was sent as a gift. (Good job it was harmless, then.) Brief Googling identified the Mumbai firm as a travel agent that offered honeymoons and tour packages. Not a promising start.

Inside the envelope were blister packs of pills, stamped with Modvigil – a brand name for modafinil – and the name and address of two more Mumbai firms. Both called themselves pharmaceutical companies and more searches on the internet suggested both seemed to check out as outfits that, if they wished, had the knowledge and facilities to knock out a decent batch of modafinil.

Each blister pack was stamped with a batch number, their claimed date of production (January 2015) and expiry date (December 2017). To most buyers, I suspect, that would seem pretty convincing. But the proof of the pudding of course is in the eating, and I wasn't ready to eat one of these little white discs just yet.

Whether pills bought over the internet are genuine is one of the most common questions on the dozens of websites dedicated to discussions of smart drugs. For modafinil, there isn't an easy way to check. Dropping them in vinegar and looking for bubbles was suggested by some web-users, but it didn't add up as a way to detect the drug given the range of possible additives in there. Genuine modafinil, plenty of users pointed out, would make my wee smell foul, but that didn't sound scientific.

I needed professional help, but the professionals seemed unwilling to offer any. I asked big contract testing labs who said they didn't deal with individuals. I thought about rebranding myself as a honeymoon tour company that sent gifts of harmless medicines halfway across the world, but decided it would be simpler to ask friendly chemists who worked in universities. They would be interested in the result as well, I reasoned. Although there were plenty of warnings from academics about students risking their health taking unchecked imported drugs, nobody I could find had actually, well, checked them.

At first my emails were politely rebuffed, passed to colleagues or simply ignored. 'An interesting question but not one I can help with,' said a typical reply. Then after a couple of months of discussions I finally got a bite: an enterprising university department that sold time on its equipment to top up its research funds was willing to perform some simple tests. But before the deal was sealed, I had to convince the university press officer that, no, I wasn't writing an exposé on the use of such drugs among her students. And I had to give the department a blank cheque. They would spend as little as possible, they promised. And I had to send them my Indian modafinil. Technically, I suppose, if the modafinil was real, this made me an illegal supplier. Not that I expected the university scientists, after all this, to eat it.

For £230 I got an hour or so from a skilled technician who put my claimed modafinil in a mass spectrometer and subjected it to single-crystal X-ray diffraction. The results showed it was 'beyond doubt' genuine, they said. And they found enough of the drug in one of the tablets to confirm it wasn't cut or laced with anything else. The pills, as far as science could tell, were legitimate. The university now turned illegal supplier and returned the rest of the pills to me. They dropped back onto the mat.

I took my first modafinil pill at eight in the morning. I sent in some cereal and toast first and then swallowed the little white cross-headed lozenge with plenty of water. Two hours later I was sat in my usual coffee shop, writing on my laptop (working on this book), waiting to feel different. I had tried to avoid reading too much about other people's modafinil experiences because I didn't want to seed ideas but I did read that it can take a couple of hours for the effects to kick in. And then they can last for sixteen hours plus. I didn't fancy being awake much past midnight, but I didn't want to get up and have breakfast any earlier.

Having said that, I did feel different: capital letters different. I felt good, like I was concentrating on the words I wrote in a more deliberate way. I felt a connection to the writing and the screen of my laptop. The music (Christmas songs on strict rotation), other people and the kids running around were less of a distraction. I was thinking these sentences as fast as I could tap them out. I came to this coffee shop because they offered free refills. But after thirty minutes my first cup lay pushed to one side, barely sipped and cold.

Was I making this sensation up, imagining it? Was this just a placebo effect, the power of suggestion? Did it even matter? I had taken a drug supposed to sharpen my senses and release my cognition, and my senses felt sharp and my cognition free. I felt like I wanted to keep typing. So I did. I could FOCUS and I felt MOTIVATED.

In previous days two hours of book writing was about enough for one sitting. I would kid myself I was still making progress beyond that, but my attention would wander and the writing slow. That wasn't happening that day. Several hours in and I was still alert. The screen seemed bigger and more welcoming. I felt like I was leaning in, the words as they presented themselves seemed to be close and moving smoothly and quickly. This was *terrific*. If it was a placebo effect then bring it on.

If pilots fly helicopters and fighter aircraft on these drugs then I'm not sure if that's a good idea. Alongside the welcome sharpening of my senses I felt impulsive and my fingers were twitching when they weren't striking the keys. I stroked my unshaven chin a lot. I didn't think I would like to drive. I hadn't spoken or stood for two hours. I felt like I didn't want to. My head felt like it was where it was all happening.

I took the modafinil on a Tuesday because I usually play squash on Tuesday nights and I usually lose. I play against my friend Mike and have done for years. If you've never played squash then I recommend the book *Saturday* by Ian McEwan for a first-class description.

It's an intense, committed sport and, unlike tennis, you compete for territory and court position directly with your opponent. It's personal. You feel the court shake as they run past you. Mike and I push each other out of the way as we go to hit the ball.

One reason why Mike usually wins is because of something sports psychologists call TCUP. Thinking Clearly Under Pressure. Hot, bothered and frustrated, I lose concentration and start to slash at the ball and play shots I know immediately are wrong. I make mental mistakes. More importantly, I make more mental mistakes than Mike does. The other reason is motivation. He hates to lose more than I love to win. I would happily lose 3:2 in a tightly fought match rather than crush him 3:0. He wouldn't. (He would also point out, fairly, I am not in a position to judge, having never crushed him 3:0.) Modafinil should help me cut out the mental mistakes, which is why it is banned in competitive sport.

Technically, tonight I will be a drugs cheat, a doper, a fraud who will devalue the honest spirit and purity of sportsmanship. But I am willing to sacrifice my sporting soul for scientific enquiry. If modafinil gives me the extra mental edge I need to beat Mike at squash then it really is a smart drug. And I might take another one on Friday. I'm supposed to be playing golf with Jim.

That evening I returned home to the familiar feeling of defeat. Bugger. I'd lost. Again. And yet . . . For a while I was brilliant. I made correct decisions. I felt focused. I felt GOOD. I'd done nothing special, just put the ball in the right places, down the line and tight to the wall, resisting the ambitious and letting him make the mistakes. I won the first game.

In the second game I played one of the best shots of my life. Not that Mike noticed, or anyone watching would have been impressed. He hit a weak service return, high on the front wall and the ball came looping towards me. I set up to pounce and to smash it low into the front right corner. It's a tricky shot to get right – hit the ball too hard in squash and it just bounces further away from the wall and makes it easier for the opponent. Hit the jaws, where the front and the side wall meet, and it squirms back towards the middle where he can put it away. It's a shot I typically mess up, but I felt like I should try it anyway.

Except this time I didn't. I saw the future and I changed it. I thought clearly under pressure and chose a different, safer shot. Instead of angling the ball down from above my head I shifted the head of the racket and pushed it through the ball, which went up, not down, and arced high over my left shoulder and dropped into the left-hand back corner. Mike was as surprised as me. Anticipating the smash he had darted forwards and could only watch, wrong-footed, as the ball drifted beyond his reach. I won the second game. As Mike, exasperated, reached for his water bottle he muttered to himself and then yelled a single word in anger – 'CONCENTRATE'. I allowed myself a little smile.

I won the first three points of game three. Incredible. I had won the odd match against Mike, but never 3:0. What a result that would be. What a story. What a METAPHOR. I felt like Malcolm Gladwell, the Canadian journalist who writes bestselling books about simple solutions to complicated problems. What an opening scene to my book this would be. People wouldn't believe me of course, so I would have to ask Mike to send an email confirming what had

happened and how amazed he had been at my cool, clinical, focused play that night. Then I could publish it as a
footnote. No, an appendix.

'6:3.'

What?

'The score. It's 6:3 to me,' said Mike as he served. I tried
a clever return, playing it off the side wall to die in the front
corner. Risky but he would never expect it. The ball hit the
line. Out.

'7:3.'

Whatever I had, the mental boost of a banned stimulant,
the confidence of the placebo effect or just the increased
concentration from thinking so much about the mental
side of the game, it had gone. The bad old me returned to
court. The TCUP well and truly dropped. Mike won the
next three games and the match. He won and I lost 3:2 in
a thriller, so I suppose we both got what we wanted.

Those first two games were so strange, Mike said to me
in the pub later. I just couldn't get going and I kept making
the wrong decisions. You didn't seem to make any mistakes. I nearly told him the truth but I chose a different,
safer shot and stuffed a chip into my mouth instead. Next
Tuesday, I thought, I would take the modafinil an hour
later.

Don't judge me. Between 1984 and 2004, many of us
could have been drug cheats. Included on the WADA list
of banned stimulants then was caffeine, and two Olympic
athletes were even caught and punished for using it. The
Mongolian judo star Bakaava Buidaa was stripped of the

silver medal he won at the Munich games in 1972 for excessive caffeine intake, and Australian modern pentathlete Alex Watson was thrown out of the 1988 event in Seoul. (He later cleared his name and competed in Barcelona in 1992.)

Recreational use of caffeine – the odd cup of coffee – was ok, but once levels rose above a certain threshold, sporting officials assumed someone was trying to gain an unfair advantage. The cut-off point was high, but it wasn't *that* high. About six cups of strong coffee could put someone in the danger zone.

Caffeine has been used for centuries to keep people alert. The writers Voltaire and Balzac are said to have drunk dozens of cups of coffee a day. In my university days in the 1990s, caffeine came in concentrated tablets, but they were pretty weak: each contained 50mg of caffeine, about half of a strong cup of coffee. Students today can call on much bigger guns, and caffeine tablets with 200mg are available. (In Germany they are classified as a controlled medicine for fatigue and called, really, Coffeinum.)

Doctors recommend a maximum of about 400mg caffeine a day (less than the 500-odd found in the largest cup of Starbucks) but they occasionally meet people who have taken much more. A forty-two-year-old in Ohio swallowed 120 caffeine tablets (each 200mg) in a suicide attempt. He survived, but only after four days of uncontrollable vomiting and diarrhoea and slipping in and out of consciousness. He had a record 24g (24,000mg) caffeine in his system.

The best guess of scientists is, when it comes to restoring

alertness, modafinil packs more of a punch per mg than caffeine. About 400mg of modafinil does the same job as 600mg of caffeine. That would make my pill with its 200mg of modafinil about the same as necking three cups of strong black coffee at once, or taking six Pro Plus.

There is an important difference though. While caffeine is classed as a mild cognitive enhancer, and has been shown to sharpen reaction times, it has almost all its impact on making tired people feel normal, rather than making normal people feel super. Nicotine does the same. Modafinil is different. While coffee can help tired people feel more alert, modafinil seems to stop them feeling tired in the first place. It's not just about bringing people back to normal; it seems to have the ability to take them beyond.

The effects of smart drugs are often hyped and exaggerated. But solid evidence suggests modafinil has a positive and significant effect on cognition. It's been shown to improve the performance of healthy volunteers in several tasks – recalling a series of numbers, decision making, problem-solving and spatial planning among them. In August 2015, scientists at Harvard and Oxford universities pooled and analysed all of the most reliable experiments and concluded modafinil is the world's first safe and effective smart drug.

By safe, they mean in the short-term. Nobody knows what the long-term effects might be, partly because scientists haven't tracked chronic modafinil use, and partly because they are not sure how the drug works, or indeed what it does in the human brain. It's notoriously hard to track the actions of medicines inside the brain because, unlike the other organs, it's difficult to tell much about what goes on in the brain from conventional tests like blood samples. The brain is bathed in its own syrupy fluid, which is installed in a separate circulatory system to the main blood supply and kept isolated by the blood-brain barrier.

The only real way to directly check on levels of chemicals and drugs and how they might change in this cerebrospinal fluid is to hack into the plumbing lower down the body, and these lumbar punctures (spinal taps) are risky and so are never done lightly. That rules them out for research studies on how well smart drugs can make university students recall strings of numbers.

From experiments with animals and cell cultures, and from studying brain scans, neuroscientists think modafinil probably changes the activity of neurotransmitters, which help direct brain activity by helping the separate neurons communicate. Specifically, modafinil seems to affect the catecholamine system, which produces and releases the neurotransmitters dopamine and norepinephrine. This might concentrate activity in parts of the frontal cortex involved in higher mental function, while inhibiting it in neighbouring regions, so reducing competition for the needed cognitive resource. (That is certainly how it feels – modafinil helps keep the brain on the task at hand and immune from distraction.)

Some scientists are sceptical that modafinil makes the cognitive gears turn faster. It might simply increase attention and motivation. Having taken the drug I can see their point. I could probably attribute many of the effects I felt to heightened motivation, or at least a greatly reduced desire to go and do something else instead.

But if what we want to measure as intelligence is the output of someone's brain, rather than the internal work-ings, then isn't the point moot? As we discussed earlier, we don't take nerves or a lack of confidence into account when we mark IQ tests or exams. And couldn't those have as big a (negative) impact on someone's scores as increased moti-vation could see them improve?

Although modafinil is widely regarded as safe, some people do seem to react badly to it. Particularly worrying for me, before I took the drug, I found reports that two patients with OCD – people who had been treated success-fully and had in effect beaten the condition into remission for at least a year – dramatically relapsed when they took modafinil. The psychiatrists couldn't be sure what was going on; maybe it caused a part of their brains involved with obsessions to flare up again.

And in 2015, psychiatrists in Turkey reported a patient prescribed modafinil for excessive sleepiness – she felt like she had to go back to bed for an hour or two each afternoon – developed hypersexuality. She still wanted to go to bed each day, just not to sleep. Married with two children, the forty-five-year-old found her massively increased sexual desire a problem. So did her husband, who was seventy-five.

Modafinil, I can report, did not have either of these effects on me.

Beyond modafanil, there are other medicines that healthy people use to increase their cognitive performance, and more are on the way. Medical amphetamines including Benzedrine have been on the market for decades (and were given to RAF pilots in the 1940s) while newer drugs, such as the Alzheimer's treatment donepezil, are being developed to address the looming crisis in dementia. Among the most common of the so-called study drugs is Ritalin, prescribed (many say over-prescribed) for children and others diagnosed with Attention Deficit Hyperactivity Disorder (ADHD). It helps them focus and stay alert. As such, it too is banned in competitive sport. It's outlawed in major league baseball in the United States unless a baseball player has been diagnosed with ADHD, in which case he is granted a therapeutic exemption. The result is an epidemic of ADHD among baseball players, who have a diagnosis rate for the condition twice as high as the rest of the population.

Cognitive enhancers are popular among amateur athletes too. An anonymous survey of almost 3,000 competitors at triathlon events in Germany found 13 per cent admitted to physical doping in the previous twelve months, while 15 per cent said they had experimented with cognitive doping. (Because of the legal status of caffeine tablets in the country, those counted.)

It's not just the athletes themselves who see the appeal of smart drugs. A 2016 editorial in the *British Journal of*

Sports Medicine argued team managers and coaches, who are increasingly expected to rely on complex statistics and make rapid decisions based on large amounts of information, could benefit from brain doping and, as this would give their teams an unfair advantage, these non-playing staff should be drug tested too.

Smart drugs are actually pretty dumb. They saturate the brain with active ingredients and hope that some manage to find a suitable target. It's impossible to use them to target a specific brain region and so a specific function – say memory or problem-solving. To do that, we need to zoom in a little, to break down the whole brain into its constituent parts and then identify the bits we are most interested in, those that control and produce intelligence.

That can't be done with drugs. It takes a more focused and hands-on approach. And, in the search for the secrets of intelligence, this is a common strategy. In fact, intelligence researchers have been encouraging a hands-on approach for an awfully long time.

SIX

The Mutual Autopsy Society

In 1892, the popular US poet Walt Whitman died and left his brain to science. It's a good job he wasn't around to see what happened next. Science dropped it. Whitman's brain, the source of some of America's favourite verse, hit the floor and broke into pieces. This probably wasn't what Whitman had in mind when he wrote the famous line, 'If you want me again, look for me under your boot soles.'

No matter, there were still plenty of good brains to go around. This was the age of the gentleman scientist, and in the late nineteenth century there was nothing that marked a scientist as a gentleman quite as much as his willingness to allow friends to have a good rummage around in his head when he passed away.

Whitman had hoped for greater things. He had asked for his brain to be removed as part of a worldwide scientific effort to locate the anatomical basis for intelligence. These early neuroscientists were looking for markers of intelligence in the brain, and to do so they made a simple assumption: bigger is better.

It makes sense a larger brain would indicate more intelligence. Your brain accounts for about 2 per cent of your body weight yet demands 20 per cent of the oxygen you breathe. A fifth of your food goes into powering the brain and its billions of cells. The more brain cells, the more they can do, and the increasing size of the brain during human evolution is linked to the development of more complex, intelligent behaviour. We scoff at the dinosaurs because we hear their brains were the size of walnuts. (In fact dinosaur brains were a decent size.)

Just like early ways to analyse IQ, the inspiration for Walt Whitman and his friends to measure and compare brain size to find the source of high intelligence had originated in France, where a group of scholars, academics and committed secularists in Paris formed the brilliantly named Mutual Autopsy Society. Each member pledged that on his death, the others could hack open his skull, retrieve his fresh brain and place it on display for the public.

In death, these members of the Mutual Autopsy Society hoped to make a point they failed to prove in life: there was no soul and so, contrary to religious teaching, humans did not deserve to be placed on a higher spiritual plane than any other animal. New members to the society would pledge their allegiance with a solemn oath. 'Free thinker, loyal to scientific materialism and the radical Republic, I intend to die without the interference of any priest or church.'

Similar brain donor clubs cropped up in Russia, Germany and Sweden. But it was in the United States the idea really took hold. Unlike those across the Atlantic in France, the God-fearing men of America did not want to prove the non-existence of a higher power. They wanted to demonstrate they – and their esteemed colleagues – were *themselves* a higher power. They wanted to use the size and shape of their dead brains to prove their kind were more intelligent than the rest.

More than a century on, we have scanners today to watch a brain at work. But much of what we know about the brain function still comes from the kinds of natural experiments carried out in Whitman's day, which observed the impact of brain injury and disease. The most influential was work back in the 1860s, when the neuroscientist Paul Broca, with the help of stroke patients who had lost the ability to talk, pinned down generation and control of speech to the frontal lobe part of the brain where the damage was concentrated – now called Broca's area.

Could the control of cognitive ability – the seat of intelligence – be found also? Some scientists thought so, and a school of research known as phrenology identified intellect as a key human trait divined by assessing the physical bumps and lumps on the surface of the skull. Greater intellect, the phrenologists argued, would swell that region of the brain on the inside and the increase would show up on the outside. Some popular terms we still use to describe intelligence come from this period. Highbrow, for instance, was originally a physical description, because the phrenologists associated a high forehead with cleverness (lowbrow was the opposite). Telling someone to get their head examined was first an invitation to visit not a psychiatrist as we would say today, but a phrenologist.

As the phrenologists fell from fashion, the search for intelligence switched from the outside of the skull to the inside. A new generation of researchers worked with dead bodies. At first they threw away the brains. They boiled empty skulls clean and plugged eye sockets with rags and cloth. To measure the size of the discarded brain, and by inference the cleverness of its owner, they stuffed the space it had occupied with water, mustard seeds or lead shot, then tipped the contents out and measured them. Skulls were easy to collect and keep. Collections built up, sometimes hundreds strong.

Although these skulls were measured in the name of science, using them to search for intelligence was a cover for darker motives. In most cases, skull collections were used to support claims of difference between races. More accurately, they were used by white men to supposedly show how other races were inferior.

Among the keenest of these collectors was the anthropologist Samuel George Morton, who gathered and measured more than a thousand skulls from across the world, including bones from South Africa and Australia. Morton claimed white people had consistently bigger spaces in their heads for brains than black people. This fed into the belief at the time that whites and blacks were different species, and the whites, because of their extra brains, were superior.

The heads and skulls used in these comparisons were

anonymous, often scavenged from battlefields, and this limited how they could be used. Beyond ethnicity, the remains said nothing about what the dead person had been like in life – what they had done and how intelligent they had been.

To prove larger brains produced more intelligence, these early scientists needed to go a step further. They had to connect the larger heads and brains they measured to the great abilities and achievements of their former owners. This is where Walt Whitman and his friends saw an opportunity.

Inspired by France and the Mutual Autopsy Society, a group of self-regarding men of the northeast United States formed what came to be known as the Brain Club. They preferred the grander title the American Anthropometric Society. Similar to the French, each man pledged the others could remove his brain after death and examine it for clues to his greatness.

Up to 300 men are believed to have joined the society, but few admitted it publicly, and even fewer went to the trouble of writing it into their will. Walt Whitman never did. Historians think the poet – long fascinated by the brain and friends with many of those who studied it – probably agreed to donate his brain but didn't tell his family. Certainly his brother, George Whitman, was horrified at the idea and, when Walt did die, did not want there to be an autopsy.

The idea of Brain Club members was simple: it all came down to size. Bigger, heavier brains, they reasoned, held more potential and more ability, and so should confer upon their owner more status. As they removed and weighed each other's brains, they convinced themselves the idea was correct. They published league tables of brain weight, with the heaviest brains of their professional friends and colleagues – physicists, lawyers, composers, humorists, mathematicians, politicians, economists, editors, writers, geologists and judges – grouped towards the top. The undisputed brain heavyweight champion was the Russian poet and novelist Ivan Turgenev, who left behind a brain of 2,012g, the first and only to break the 2kg barrier.

The reverse was also true, the Brain Club believed. People of lower status – bricklayers, blacksmiths and labourers – had smaller and less powerful brains, they said, and appeared in mid-table.

At the bottom of the league were those people given the smallest and stunted brains, as incapable of producing intelligence as the owner was of moral and intellectual achievement. These people were the criminals, and there were plenty of their brains to go around.

One of the most high-profile criminal brains from the period was taken from the anarchist Leon Czolgosz, who shot US President William McKinley near Niagara Falls as the two men went to shake hands. It took President McKinley more than a week to die from his wounds. By the end of the following month, Czolgosz was dead also, after being quickly tried, convicted and then executed in the electric chair.

Within an hour of his death, Czolgosz was in pieces on

the post-mortem slab. His brain was the prize, and, surprisingly, given the high profile of the case, it was removed and described by a fourth year medical student. The brain was normal, disappointingly so for those who believed criminal tendencies would show up not just in small size but as physical features. 'It is a probable fact that certain oft-mentioned aberrations from the normal standard of brain structure are commonly encountered in some criminal or degraded classes of society,' the young student wrote in his autopsy report. 'But these structural abnormalities, so far as they have been described in the brains of criminals, are too few and too insufficiently corroborated to warrant us drawing conclusions from them.'

The medical student concluded the assassin was socially diseased and perverted, but not mentally diseased. 'The wild beast slumbers in us all. It is not always necessary to invoke insanity to explain its awakening.'

The student's name was Edward Anthony Spitzka, and his career was nearly over before it began, when an unscrupulous stenographer who transcribed Spitzka's words during Czolgosz's post mortem tried to sell them to the press. The student was forced to write to medical journals to warn them that if they published a 'garbled rendition' he would disclaim it.

Spitzka perhaps got the job of cutting open a presidential assassin despite his inexperience because someone knew his father. Edward Anthony Spitzka's father was called Edward Charles Spitzka, and he forged a similar career to the one his son would pursue, in neurology and implications for society.

Most notoriously, Spitzka Senior had testified in 1881 that another Presidential assassin, Charles Guiteau, the killer of James Garfield, was insane. In a bad-tempered appearance in court, Spitzka Senior was forced to deny prosecution charges that an academic post at Columbia Veterinary College meant he was not a psychiatrist.

'You are a veterinary surgeon, are you not?' he was asked.

'In the sense that I treat asses who ask me stupid questions I am,' he snapped back. Despite the testimony, Guiteau was found guilty and hanged.

Spitzka Senior was present at the execution of William Kemmler in the electric chair – the story that begins this book. And he was an original member of the US Brain Club, passing control of the society to his son in 1902. When he did so, Spitzka Junior found his dad's prized assets in a dreadful state. Three of the founder members of the society had died by then, but two of their stored brains had flattened and become distorted. Of the donated brains of other members, at least two were in pieces, one because it had been left to float in hardening fluid for ten years. Walt Whitman's, of course, was missing.

Spitzka Junior took studies of brains for signs of intelligence and esteem out of the shadows. Buoyed by his well-received analysis of the executed Czolgosz, he investigated more brains, both of criminals and the great and good. (When his father died it was Spitzka Junior who removed and measured Spitzka Senior's brain.)

As he chopped skulls and analysed brains, the younger man was explicit in his scientific goals. 'It is not enough merely to admire the genius of an Archimedes or a Homer, a Michelangelo or a Newton; we wish to know how such men of brains were capable of these great efforts of the intellect.' Given so many great men were willing to leave their brains to science, he added: 'It is our business to endeavour to ascertain why and how some are more, some less, gifted than others.'

These investigations of brains were crude and messy and unreliable. Done in a proper scientific way, the researchers should have not known if a brain being examined had belonged to an esteemed colleague or a common criminal. These early scientists were doing it the other way around. They knew whose brain they were measuring, and given their idea that successful men had larger brains, it's not surprising they measured them in that way, because they wanted them to be so.

Anomalous results were discarded or explained away. An unusually light brain of a great man was excused, and said to be down to the degradation of ageing, or because bits must have been left behind when a clumsy technician scooped it from his great head. The too-heavy brain of a lesser individual was blamed on disease or the chemicals used to preserve it. The data were massaged to fit the pattern and the world order the scientists believed in. This form of cognitive bias is a common trap for scientists, and the early anthropologists were far from the first, or last, to fall into it as they pursued the mystery of intelligence.

Spitzka Junior's own studies of executed criminals helped convince him an unusually heavy brain in the less gifted could be explained by disease or abnormality. It was unfair, he said, to include the weights of these brains in any true scientific analysis. 'Those great water-logged pulpy masses in the balloon-like heads of hydrocephalic idiots did not discover and never could have discovered the laws of gravity, invent the ophthalmoscope, create *Hamlet*, or found modern natural history.'

He added: 'The brains with which we here concern ourselves are those of men with healthy minds who, in their life time, attained high distinction in some branch of the professions, arts, or sciences, or who have been noted for their energetic and successful participation in human affairs.'

Their efforts sound crude, but more rigorous studies and endeavours of modern neuroscience do confirm these early intelligence researchers were on to something. Large brain size and greater IQ *are* linked. It's not a massive effect, but it is significant. The same goes for head size, presumably because large brains need large heads to hold them. As crude as it sounds, the simplest way to gauge someone's mental prowess is a tape measure around their head.

In 2007, scientists in Edinburgh used head measurements to estimate the intelligence of Scottish national hero Robert the Bruce (victor over the English at the 1314 Battle of Bannockburn). They analysed a cast of his skull prepared when Bruce's body was exhumed in 1819.* Bruce, the scientists said, had an IQ of 128 and maybe higher. That's about right, they claim, for a man behind the 1320 Declaration of Arbroath, which proclaimed Scotland as free from English rule and is credited by some historians as the inspiration for the US's own Declaration of Independence.

The confirmed link between skull size and intelligence would no doubt please the members of the Brain Club and the Mutual Autopsy Society, and it shows they were on the right track. But the link doesn't help when it comes to cognitive enhancement. We don't have a way to make our heads and brains bigger and nor are we likely to in the future.

To find ways to boost the workings of the brain, we need to be more sophisticated and look inside. Could the shape and structure of the brain perhaps offer an insight to the source of intelligence? If so, then it should show up in the brain of a man whose name has become shorthand for genius.

The strange story of what happened to Albert Einstein's brain after his death has been told many times. But it's still worth recording here some highlights, if only to demonstrate the continuing allure the secrets of intelligence have

---

* Bruce's body never really rested in peace. His heart was removed on his death and taken on the Crusades against the Moors in Spain. Returned to Scotland, it has been dug up and reburied at least twice more.

for modern scientists; secrets that Albert has been reluctant to reveal.

Einstein knew his brain would be targeted. And unlike the members of the Mutual Autopsy Society, he had no wish for it to become a laboratory exhibit. Before he died he seems to have given clear instructions: his remains were to be cremated and scattered in secret.

Yet during the 1955 autopsy into the cause of Einstein's death (a burst aorta), his brain was secretly removed by a pathologist who believed he could use it to make his name. The pathologist, Thomas Harvey, chopped it into more than two hundred pieces and prepared over a thousand tissue slides, each of which contained a thin slice. He posted these out across America, to seek the opinions of the leaders in the field. The rest of the brain he kept in jars in a cupboard of his Princeton University office, resisting for decades enquiries and requests to examine it, including from the US Army. If studies of the posted slides were ever carried out, the results showed nothing out of the ordinary, and the scattered pieces of Einstein's brain were left to gather dust in drawers and attics. Most are still out there.

After a journalist wrote about Harvey's work with the brain in the late 1970s, requests from scientists for new pieces to study came pouring in. Again, the enterprising pathologist popped slides in the post.

Together with detailed photographs Harvey had taken, those samples produced a wave of new studies, most of which claimed to have found something unusual. Results based on them still appear from time to time.

Einstein, according to those who have examined his brain, had an unusually high number of glial cells, which nourish neurons and keep them in place. The brain cells in his prefrontal cortex were especially tightly packed, while his inferior parietal lobule, associated with spatial and mathematical tasks, was unusually wide. As recently as 2012, new research claimed Einstein's brain had an extra ridge on its mid-frontal lobe, a region linked to planning and memory.

But in many ways, Einstein's brain was unremarkable. It weighed a pretty paltry 1,230g – towards the lower end of the normal range for a man in his seventies.

When it comes to intelligence, only so much can be gleaned from dead brains, however big and famous their former owners, which is why scans of the living insides of people's heads prove so alluring to modern neuroscience. Usually taken with magnetic resonance imaging (MRI) machines, these scans offer an eyewitness account of how parts of the brain demand more blood when their owners perform mental tasks. That's usually taken as a proxy for increased activity, and neuroscientists then try to deduce which parts of the brain are involved in, and perhaps responsible for, mental traits from cognitive skills and emotions to decision making and memory.

Charles Spearman's general intelligence, 'g', can't be found in the brain, or at least it can't be located on a scan in a specific part of the brain. It's real, but it doesn't exist in a structure that can be pointed to. It's more a measure of what the brain does; just as athletic ability is a genuine measure of how physical prowess differs between individuals, but couldn't be traced in a scan of muscles.

If we break intelligence down into some of the constituent parts – memory, maths, language, reasoning etc – then it becomes a little easier to place each of them inside the brain. Regions in the parietal lobe are known to help us identify and process visual imagery, and the hippocampus is strongly associated with memory. But while brain scan studies continue to ascribe functions to an increasing number of specialist parts, they don't explain why one person's works better than someone else's.

The brain has two types of tissue. Grey matter does the bulk of the work. White matter holds the grey in place and passes signals between different brain areas. Both seem relevant to intelligence. Just like brain volume, a larger overall amount of grey tissue seems to relate to higher intelligence, particularly so in areas including the prefrontal cortex. The same seems to be true for white, connecting matter, though the conclusion is not so clear cut. What does seem crucial is integrity of the white tissue, which makes sense given its job. Damaged connections will clearly interfere with how well the brain can work. (The progressive loss of white matter connections could explain why many cognitive abilities decrease with age.)

Some studies find people who are skilled in a specific mental ability show a measurable difference in brain structure. Most famously, neuroscientists in London reported in 2000 that London taxi drivers, who must show an encyclopaedic knowledge of the city streets to get their licence, have more grey matter than usual in the hippocampus.

While that might demonstrate that repeated use and practice of a set of mental skills can grow a specific brain region, the conclusion doesn't really work the other way around: finding an enlarged hippocampus in a plumber from Aberdeen wouldn't guarantee she could tell you the quickest route to drive from London Bridge to King's Cross Station.

These structural characteristics of more intelligent brains can help pin down the neural basis of cognitive ability, but they are no more use than brain size when it comes to cognitive enhancement. We can't go in and add grey tissue. If we want to improve the way a brain works, then we must look beyond structure and try to improve its function. So how does a more intelligent brain function?

Rather than being a product of a specific brain region, general intelligence seems to come from how effectively various brain regions can work together. To solve a problem, parts of the temporal and occipital lobes, at the base and back of the brain, first take the raw signals that flood in from the eyes and ears and process them. This information is fed into the parietal cortex, a broad arch of brain tissue just under the crown, where it is annotated and labelled with meaning. It then goes forwards to regions of the prefrontal cortex, sitting behind the forehead, which manipulate it, package it into possible ideas or solutions, and test them. As one solution emerges as preferred, another part of this prefrontal cortex, the anterior cingulate, is recruited to block the other, incorrect, responses.

Because most of the intellectual heavy lifting in that series of brain functions takes place after the processed sensory information gets shunted from the back towards the front of the brain, this model of intelligence is called the Parieto-Frontal Integration Theory (P-FIT). The better this P-FIT circuit works, then the more general intelligence a brain, and so a person, will have.

So, and we are nearing more promising cognitive enhancement opportunities here, how does one person's P-FIT circuitry work better than another's? And can it be artificially improved?

Like a computer, raw processing speed seems to be important. One way to study functional differences between brains is to monitor their neuronal activity. When each neuron fires, as it is recruited to help solve a problem or to transmit a signal, it produces a little burst of electrical current. Add millions, maybe billions, of these tiny bursts together, as happens when the brain does something, and the overall electrical buzz can be measured. The technique – EEG, for electroencephalogram – is pretty common, and involves sensitive electrodes placed against the scalp to listen for electrical changes in voltage.

EEG can track these voltage changes to investigate everything from sleep to epilepsy and it works on a simple principle: when the brain is active, its electrical activity increases. EEGs, for example, reveal more spikes when the brain works on a mathematical puzzle than when it is asleep.

Of particular interest to intelligence scientists is the way

114

the EEG can record the brain's response to a stimulus, such as a sound. Within one-tenth of a second, the EEG trace of brain activity shows a tell-tale response. There's a small dip and then, about another tenth of a second later, it recovers. The most significant action comes after another tenth of a second – three-tenths in all after – when it shows a sharp spike. That is called the brain's P300 response.

The P300 response is a hot area of research in neuroscience. Some scientists think it could offer a reliable way to spot when someone is lying. And psychologists have linked it to intelligence. Specifically, they have found the P300 response comes slightly earlier in people with higher mental ability (the difference is perhaps a few thousands of a second). Clever people seem to have a faster electrical response. And some studies link better performance on tests of intelligence to a higher P300 peak.

The shape of the three responses on the EEG chart might differ according to intelligence as well. Some studies suggest lower cognitive performance is associated with a less defined, less complex, response. The three bumps are not so obvious. Because the more complex traces associated with higher intelligence would, if straightened out, form a longer line, some psychologists call this the piece-of-string test. How long is a piece of string? It could depend on how bright you are.

A more significant difference is visible in the way clever people fuel their brain activity. Brain scans of the way glucose is used to release energy, another proxy of mental activity, show, as would be expected, that energy demand increases when the brain is put to work. In people who score well on intelligence tests, the required increase is smaller. High intelligence is linked to efficiency. Those with less effective brains need to burn more glucose to fire more neurons to solve the same problem. This could indicate more intelligent people need to recruit fewer neurons and set into action a smaller number of brain circuits.

We don't have all the answers of how intelligence shows itself in brain activity yet – analysis of brain circuitry is a new focus for neuroscience. But we do know that intelligence circuits, like all those in the brain, rely on two types of communication: chemical and electrical. And, as we'll see, neuroscience now has tools that can alter both.

In 2015, neuroscientists showed the way these brain circuits activate is highly personal. Although we all use the brain's P-FIT system to reason and problem-solve, we each do it in a slightly different way, recruiting a different number of neurons and in a different order. In fact, the neuroscientists, from Yale University in the US, found patterns of brain activity so personal they served as a kind of neuronal fingerprint. The scientists could pick out and identify people from a large group of volunteers by mapping and then looking for their tell-tale patterns of brain connections as they performed cognitive exercises.

What's more, the neuroscientists found these brain fingerprints also indicated a person's intelligence. A computer could compare scans of people of known intelligence and pick out brain connections and patterns they had in common. Then it could use that information to accurately estimate the intelligence of people it had never seen before, based on a scan of the way their brain was wired. Who needs IQ tests? In future all it might take to find the brightest in society is a scan of their brain circuitry in action.

What determines the layout and workings of these brain circuits, the equipment and infrastructure of our P-FIT thinking and reasoning system? To a large extent, like much of our physical architecture – from the shape of our nose to the colour of our eyes – our brain wiring is genetically determined, influenced by those who went before. Your brain is like the brains of your parents, and like the one you will pass on to your own children. You don't truly own a brain. You look after it for the next generation. It's a simple principle, but also a dangerous one that appeals to the worst of human nature.

In my day job, I write editorials for the science journal *Nature*. The articles tend to be aimed at a specialist audience, who work in research or are involved with the funding and support for such research. Sometimes we tackle the big issues in broader society – the refugee crisis in Europe in the late summer of 2015 was one I was quite proud of. The best editorials to write are those when the narrow interests of science and the broader issues of society overlap, on topics like climate change, new biological techniques that could breed designer babies, and so on.

Nature has published since 1869, and it's as much a journal of record as a weekly magazine to inform and entertain. Most decent libraries have bound volumes that go back decades and articles from back issues are still referred to and discussed. The refugee editorial we ran in 2015, for instance, leaned heavily on a similar piece, to address a similar crisis, which *Nature* published in 1939. The position that *Nature* takes on the big questions of the day tends to be in line with the attitudes of most professional scientists: humanitarian and evidence-based. Sometimes though, I look at editorials from past issues and wonder just what in God's name we were thinking.

In February 1926, a predecessor who held the same job I do now wrote an editorial on the subject of intelligence in *Nature*. It was titled 'Racial Purification'. And yes, it was as bad as it sounds. Trigger warning: this is where the story of intelligence takes a very distressing turn.

SEVEN

Born with Brains

Does intelligence flow from parent to child through the genes? The hundreds of women who paid Robert Klark Graham $50 for his sperm certainly hoped so. During the 1980s and 1990s, Graham collected sperm from a register of Nobel Prize winners and other high intellectual achievers and sold it through what he named the Repository for Germinal Choice. Most people called it the Genius Sperm Bank.

More than two hundred babies were born before Graham's repository closed its doors shortly after his death in 1997 (he slipped and banged his head in the bathroom at a science conference where he was trying to recruit donors). Of those children who have come forward since, some – but far from all – say they are highly intelligent. Doron Blake, the bank's second born, said in his early twenties: 'I turned out very well, my IQ was off the charts and basically I was everything Robert Graham wanted. Throughout my life I've felt I've not had to work as hard for the level of achievement that I've reached as most of my peers did.'

119

They might not sell sperm any more, but scientists still select and work with intelligent people. The biggest list of the brightest is held by scientists at the Centre for Talented Youth at Johns Hopkins University in Baltimore, who every year screen school test results to identify gifted teenagers and encourage their development. Stars including Facebook's Mark Zuckerberg, Sergey Brin of Google and Lady Gaga have enrolled in the scheme and attended its summer schools – affectionately known as geek camps – and as a result the scientists now have records of some 1.5 million intelligent people.

The brightest of these students – who score the best marks before they are thirteen years old – are invited to join an elite project, which tracks their progress to work out what makes them so special. This Study of Exceptional Talent programme has run since the late 1970s and includes information on what the gifted students achieve as adults: prizes and competitions won, patents awarded and work published.

The psychology of intelligence – Charles Spearman's discovery of g and its codification as IQ – is controversial. But it's nothing compared with the arguments and bitterness that surround the genetics of intelligence, which is so badly tainted that many psychologists and geneticists refuse to work on it at all, and argue others should not either. The controversy goes some way to explain why most universities no longer teach the basics of intelligence in undergraduate psychology lessons – an astonishing omission given its centrality to so many human abilities and behaviours.

So, when an international group of genetics experts approached the Centre for Talented Youth in 2008 with a simple request – please put us in touch with your high achievers so we can take and analyse samples of their DNA – they probably knew they were stirring up trouble. And so it proved. The request triggered alarm bells. The academics running the scheme were not sure what to do. They were haunted, they said, by 'the ugly purposes to which claims about the genetics of intelligence had been put in the past'. Those ugly purposes go back decades. And most of us have some experience of their consequences.

The problems and the controversy began when Alfred Binet's early intelligence tests followed the Statue of Liberty from France to America, around the time of the First World War. Woodrow Wilson, US president at the time, was desperate to keep his nation out of that war. He called for neutrality 'in thought and deed' and held to that line despite the widespread outrage at home and abroad caused by the German sinking of the passenger vessel *Lusitania*, which drowned more than a thousand people, 128 Americans among them. By the time German submarines started to attack all ships they found in the Atlantic in 1917, finally forcing Wilson to declare war, the US needed to quickly organize and move out hundreds of thousands of troops.

The huge scale and rapid speed of the mobilization was unprecedented and psychologists working on IQ tests saw an opportunity. They borrowed Binet's idea, adapted the questions and ignored his caveats and appeals for caution. Rather than seeking to identify those at the bottom end of the scale to offer them help, the US psychologists were more interested in creaming off those at the top. They promoted their new intelligence tests as a critical tool for the military to screen recruits for potential, and then efficiently train and distribute an optimal mix of soldiers of different abilities across regiments.

In these early psychological tests, raw recruits were asked questions on subjects including the most prominent industry of Minneapolis (flour), to why a house was better than a tent (more comfortable). Despite common claims this was the birth of widespread IQ testing, in fact the army was sceptical of the value of the tests and largely ignored them. It was not until later that the results would be taken seriously. When they analysed the scores after the war, psychologists were shocked. US recruits – a sample of the population at large and so the platform on which the nation would seek to build its industrial future – had an average mental age of 13. An entire generation of young Americans, they concluded, was mentally retarded.

The conclusion was hopelessly wrong. As written, the army tests measured not intelligence but education. Questions asked the typical colour of garnets (red) and the name of a common soap manufacturer. Yet the damage was done.

As well as tests and expertise to promote, the psychologists now had a cause to fight. They warned whoever would listen about the dangers of a feeble-minded generation. Spooked authorities around the country and then the world started to use IQ tests more widely, including in schools, as a way to identify and separate out potentially problematic children who were identified by their low intelligence. The problem they posed, these scientists reasoned, was carried in their genes. So the solution was to stop them passing on those genes, to stop them from having offspring of their own. Most of these children are dead now, but monuments to them are everywhere.

The stretch of the A50 that links the Cheshire towns of Knutsford and Holmes Chapel is not a famous road but a friend once told me about a curious incident that happened there. A friend of hers had been driving along the A50 one winter evening when she saw a cardboard box in the centre of the road. Thinking it might cause an accident she stopped her car, opened the door and walked across to pull the box to the side.

As she moved it, one of the flaps of the box's flimsy lid fell open. Inside was a child's doll. Someone had dressed it as a clown; its glassy eyes were smeared with white make-up and its nose had been covered in what looked like red blood. The woman was glad to push the box onto the verge and return to her car. Closing its door against the gathering winter fog, she started for home again.

As she pulled away, bright light flooded the interior. A car appeared directly behind her, its headlights on full beam. Annoyed at her lack of attention to the road – the contents of the box must have troubled her more than she had realized – the woman lifted her hand in apology to the driver behind, and pressed the accelerator. As she sped up, so did the car behind. She went a little faster, and so did the other. It started to flash its lights.

The woman was annoyed at this aggressive response and, having had enough shocks for one night, signalled she would stop at an approaching lay-by to allow the impatient driver to pass. As she pulled off the road, so did the car behind. Its lights continued to flash, more rapidly now.

Fearing a road-rage attack, the woman hurriedly locked the door and was relieved to hear the whirr of the central locking confirm she was secure. Just in time – a man had jumped from the car behind and was now pulling on the handle outside and yelling, his face full in her door window.

'GET OUT OF THE CAR!'

She ignored him and stared ahead.

'PLEASE, GET OUT OF THE CAR. QUICKLY!'

Startled, she turned to look at him. He was pointing to the back seat.

'SOMEBODY IS IN THERE WITH YOU. I SAW THEM GET IN. PLEASE, GET OUT.'

'What?'

'WHEN YOU STOPPED, SOMEBODY GOT IN.'

The woman went to look behind her.

'THEY ARE IN THE BACK SEAT.'

Something stroked the back of her neck.

Unlocking the car, the woman jumped out. The man outside shone a torch into the back seat and the face of a young man smiled back. He was skinny and lying on his back. White circles were drawn around his eyes and his nose was painted red.

My friend swore the story was true, but of course it's an urban myth. In this version, the location adds piquancy because the A50 in Cheshire snakes past a centuries-old country pile called Cranage Hall. And until recently, Cranage was used as a psychiatric hospital.

I grew up in the area and we didn't call Cranage a psychiatric hospital. Cranage was a mental home, a loony bin, the place where the men in white coats – and for some reason yellow vans – would cart you away to if you did or said something odd. And exactly the kind of place a young man carrying a doll painted as a clown would escape from and climb into a stranger's car.

Cranage Hall is an expensive-looking hotel now. I called in, intending to ask staff and guests if they knew of the building's history. I hadn't been able to find anything on the hotel's website about its former use and I wondered if the owners were reluctant for those staying there to make the connection.

In fact, Andrea, the friendly woman who worked behind the bar, was happy to talk about it. A tunnel in the cellar, she said, led over a mile underground to a nearby village. It was used back in the day to deliver patients to the hospital, she explained, so families could avoid scrutiny and stigma. I asked to see it, but she said it was bricked up now.

Lots of curious visitors came to Cranage Hall, she added, drawn to the past. In truth, there was little to see, the refurbishment had seemingly erased anything distinctive; some hospital buildings had been demolished and a new extension constructed. She had a factsheet somewhere that detailed the history. She would run me off a copy.

Cranage Hall hospital was one of hundreds of psychiatric hospitals that served the UK in post-war years. Every county had at least one. Conversations at thousands of different UK schools had their own local version of the men from Cranage who would come for you. But in most cases, these places were not built as hospitals at all. They started life as prisons. Prisons to house people – its own citizens – who the British government had decided were not intelligent enough to have children.

Most of these prisons were opened between the wars, in the time of eugenics. Alarmed at the apparent widespread feeble-mindedness revealed by flawed early IQ tests, psychologists and other scientists began to demand action to preserve the intellectual quality of the population. They wanted to protect the intelligence of the human race by controlling who got to breed, and with whom. (That was one of Robert Klark Graham's goals too. He wanted to use his Repository for Germinal Choice to counter the unchecked breeding of people he considered to be unintelligent and retrograde.)

Eugenics was based on shaky science, that simple breeding could control complex traits. But it was influential and so able to cause the damage it did because it appealed to those wrestling with pressing political and societal concerns. The tragedies of the First World War left in their wake a refugee crisis, with millions of displaced people looking for sanctuary. Naturally, some were heading to places like Britain and America, sparking racial and ethnic tensions.

Broken down by ethnic background, the (flawed) results of the US Army's psychological tests carried out in the First World War appeared to show that immigrants had lower IQs, which fuelled demands for controls on their movement. That 1926 *Nature* editorial on Racial Purification at the close of the previous chapter, for example, recommended that Britain only admit immigrants who scored '25 per cent higher than the mental and physical averages of the native population'. And, given the problems with feeble-mindedness in that native population, the editorial said, the government should consider moves to sterilize them. Such drastic action, it predicted, would be 'popular with the public'.

I don't know who wrote that piece, *Nature* editorials – then and now – go unsigned. But in one respect he (and it was almost certainly a he) was right, there was public support for such measures. We are all familiar with modern public information campaigns: the adverts that urge us to eat fruit, not to smoke and to walk past the escalator and up the stairs. While *Nature* was publishing that awful editorial, the UK government was producing posters that, alongside those to urge people to brush their teeth, reminded people to 'wed wisely and help to build a better nation' and told them 'the unfit are a tax and hindrance to the fit'.

The aims of eugenics seem appalling now, but in the early twentieth century they were a popular cause in polite society and promoted widely. Francis Galton, the early intelligence test pioneer we met in Chapter Three, was a big fan. Winston Churchill toyed with them. In 1910, Churchill was home secretary in the UK government of Herbert Asquith and after he saw how US states were sterilizing mentally unfit prisoners he asked officials if Britain could follow their example.

Dr Horatio Donkin, chief medical adviser of prisons, told him the idea was 'a monument of ignorance and hopeless mental confusion' but Churchill could not shake the idea. 'I am drawn to this subject in spite of many parliamentary misgivings,' he said. 'It is bound to come someday.' Some people went further and encouraged the introduction of state-sanctioned murder, which was euphemistically called euthanasia. The author D. H. Lawrence wrote in 1908:

> If I had my way, I would build a lethal chamber as big as the Crystal Palace, with a military band playing softly and a cinematograph working brightly; then I'd go out into the back streets and the main streets and bring them all in, the sick, the halt and the maimed. I would lead them gently and they would smile me a weary thanks; and the brass band would softly bubble out the Hallelujah Chorus.

Many accounts of eugenics claim that Britain avoided laws that discriminated against people on the grounds of the perceived quality of their genetic stock and so did not intervene to stop them having children, but that's not true. Rather than sterilize people of low intelligence, Britain decided to simply segregate them, to keep the men and women (and boys and girls) physically apart. They did it in places like Cranage Hall on the A50 in Cheshire.

By the time the necessary law to make this happen reached a vote in July 1913, the parliamentary misgivings Churchill mentioned had ebbed away. Just three MPs voted against the new Mental Deficiency Bill, one of them the Liberal Josiah Wedgwood. A distant relative of both Charles Darwin and Francis Galton, and great-great-grandson of the founder of the famous pottery firm that shares his name, Wedgwood, MP for Newcastle-under Lyme, staged a one-man campaign to derail the Bill, which he dismissed as the work of 'eugenic cranks'. Over two late-night sittings of the House of Commons, fuelled with chocolate and barley water, he tabled 120 amendments and made 150 interventions in a futile attempt to block it. 'I was nearly off my head at the time,' he said later.

Once passed, the Act required local authorities to find and lock up people who could now be legally classed as feeble-minded.

How could these early eugenicists in Britain and elsewhere find their targets? How could they identify the feeble-minded menace they were so anxious about? After all, many of these claimed unintelligent people looked and behaved normally. (As they would, the eugenicists argued rather unconvincingly, as the problem was within.) On a visit to New York State Custodial Institution for Feeble-minded Women at Newark in 1905, UK experts had struggled to spot any signs of low intellect. The inmates, they noted, could 'converse reasonably'. Only when told by their hosts that many of the women were nymphomaniacs did the British visitors concede 'on close inspection' they could spot their defects after all.

Diagnosis was frequently performed by medical examiners. Here's how one described a typical consultation for possible feeble-mindedness in a young boy. He wrote:

> Imagine, if you please, a father bringing his boy at the age of eight or ten years to your institution. In the father's mind, he is a very dear little fellow, and aside from the fact that he was a little late in cutting his teeth, talking and even walking, the father can see little that is wrong with the boy.

To which the reasonable response from the twenty-first century might be: He sounds normal enough so far, Mr Medical Examiner, now what? The rest of his assessment reveals a baffling haste and lack of rigorous examination, or even understanding, of an eight-year-old child in the presence of a stranger.

> You look the boy over hurriedly. You find that he is small for his age, that he has large thick lips, with mouth open a great deal, thick tongue, abnormally large ears and a head that is rather flat and narrow through the temples, but the forehead is very prominent . . . He talks quite a little when he feels like it. But is often stubborn and will not talk at all . . . He has not been to school a day in his life, neither has he been to church or Sunday school.

His hurried examination ends with this abrupt conclusion.

> In all probability, the child will never develop, so it would be wise and to the best interest of the child, of the family, and of society, ever to discharge him from your institution . . . We must remember that many people do not believe as we do in regard to this question . . . [But] it occurs to me that no-one is better qualified from personal experience to the people at large than we.

And that was that. After the briefest of inspections and with the stroke of a pen, tens of thousands of children were diagnosed as feeble-minded. They were stripped from their families, robbed of their futures and denied any chance to prove the so-called experts wrong. There were no appeals and no second chances.

Some children were dumped at institutions by families who could not or would not look after them. Others were literally kidnapped – plucked from the streets (often, ironically for a policy to reduce the societal burden of the incapable, while they walked to and from jobs in the mills).

Some children locked up as feeble-minded did have some form of genuine mental retardation, and so they might have benefitted from the security of an asylum, but plenty didn't. Some were deaf; others were unruly or had done something to draw the displeasure of those in charge.

In 1993, a television documentary to mark the closing of the UK asylums and the dawn of the new policy of care in the community interviewed former inmates, plenty of whom explained in articulate and measured words how they had been wrongly diagnosed, sometimes out of malice, with abnormally low intelligence, and how the system ignored their pleas. The programme estimated a third of the people locked up for decades by the British government were wrongly classed as feeble-minded. That's 40,000 people.

The more I researched this book and learned about what happened, the angrier I got. I could not understand why more people today do not make more of a fuss about the injustice carried out in the name of the state, science and for the supposed benefit of us – the future generations who needed protection from these intellectual weaklings and their progeny.

Then I realized the reason why these people have no voice; no one left to speak on their behalf, to howl at the unfairness of it all. The people diagnosed as feeble-minded, if the above consultation and thousands like it can truly be called a diagnosis, and then treated so badly were, by definition, not allowed to have children. Parents, brothers and sisters have long passed away. Nieces and nephews are probably unaware, or silent about the secret shame of someone they knew only as mad Uncle Jack and crazy Auntie Jean.

Beyond the UK, other countries introduced their own policies to stop the feeble-minded from breeding. One of the first actions of the Nazi regime in Germany was to pass a 1933 eugenic law, which demanded doctors report 'unfit' patients including the mentally handicapped to special courts. It was the first step on the dreadful march to the atrocities of the Third Reich. Yet it was not a Nazi invention. They based it on a draft drawn up in 1922 by Harry Laughlin at Cold Spring Harbour in New York. The United States sterilized at least 42,616 people classed as feeble-minded between 1907 and 1944.

The rise in eugenics and the discrimination against those judged of low intelligence is often blamed on the rise in popularity of IQ tests in the early decades of the twentieth century. That's not entirely true – it was driven more by the racism and elitism of the times, and how these attitudes were used as a political lens to view, and express concern about, the mass movements of people. It did not always need an IQ test to diagnose and condemn someone as of low intelligence. But the spread of IQ tests did make these decisions appear more scientific, and gave them a veneer of legitimacy. It outsourced the reasoning for someone to be imprisoned, sterilized, or even executed away from the individuals who made the decisions and towards a numerical scale that appeared neutral and non-judgemental. Eugenics did not depend on IQ, but to achieve its widespread infamy, IQ did depend on the eugenicists. And although eugenicists did not need the IQ tests, for their ideas to prosper they did require something else. Low IQ and feeble-mindedness had to run in families. They needed genetics.

Of Shakespeare's bad guys, it is Caliban, the sub-human slave of the sorcerer Prospero in *The Tempest,* who can divide opinion the most. Despite his scheming and his attempt to rape Prospero's daughter Miranda, some insist Caliban should not even be classed a villain. Caliban, his defenders claim, is as much a victim as anyone else in the story – orphaned, captive and sensitive to the beauty and magic of the island home stolen from him.

Shakespeare certainly wanted us to see a positive side to Caliban and gave him some of the play's most memorable lines – his speech about the isle being full of noises (performed by Kenneth Branagh dressed as the engineer Isambard Kingdom Brunel) was a highlight of the ceremony to open the 2012 London Olympics.

Caliban earns sympathy largely because of his parents. His mother was a vicious witch banished from her home and his father, according to Prospero, was a demon. If Caliban was bad then he was born that way. He's an example of pure biological determinism. Prospero says as much, in a line that, as so much of Shakespeare's work does, remains acutely relevant several centuries on.

Caliban, the sorcerer says, is, 'A devil, a born devil, on whose nature nurture can never stick'.

He's a lost cause – no amount of attention, education or inspiration can save him from his innate flaws, so why even try?

To Francis Galton and the other eugenicists, determined to save the world from the peril of low intelligence and the feeble-minded, Prospero's view of Caliban would have been something of a motto. Indeed, it is probably no coincidence that Galton adapted Shakespeare's nature–nurture axis for the conflict between the circumstances of a person's birth and their environment. (Hence Galton is often credited with inventing the term 'nature versus nurture'.) Just like natural born devils, there was no hope for the less intelligent. They and their social curse had to be stopped, and that meant – the eugenicists said – they must not be allowed to have children.

That assumed, of course, feeble-minded parents would have feeble-minded children; that cognitive ability and disability would slide through the generations as easily as red hair or blue eyes. That was an assumption the eugenicists were happy to make, indeed they wrote endless books and pamphlets to make the case, and in doing so they unfairly demonized families and even entire communities.

Using the newly rediscovered work on early genetics by the monk Gregor Mendel – who crossed pea plants and then worked out the basic laws of inheritance – the eugenicists of the early twentieth century said intelligence was a trait passed on from parent to child. And on this point they were mostly right. For all of the controversy then and now over the genetics of intelligence, the basic science is pretty simple. IQ *is* a heritable trait – intelligence *does* run in the genes. If you have intelligent parents then you *are* more likely to be intelligent yourself.

The strongest evidence for this comes from studies of identical twins. In the past, such siblings given up for adoption were sometimes separated, and the babies could then grow up in different circumstances. Tracking these people down then gives scientists a powerful way to distinguish the impact of genes from the effects of environment.

Studies of intelligence in twins show those raised apart have cognitive abilities in later life more similar to each other than to members of their adopted families. Intelligence is not *all* down to genes (and even those bits that are can be held back by environmental factors like poor nutrition) but the right genes can offer a significant head start.

A major reason why this relatively simple finding is so heavily disputed, and why geneticists who want to study it draw such criticism, is because the genetics of intelligence has become entangled in another loaded social and political issue: race.

Though the gap is narrowing, various published studies show that groups of black people in the United States have scored, on average, significantly lower than groups of white people on IQ tests. (And groups of East Asians tend to outscore whites.) Entire books have been written about this finding, and every possible cause has been investigated and talked about, sometimes cautiously and sometimes less so. The one thing almost everybody agrees is that there a genuine difference to explain – in other words, as much as we might wish it away, the IQ gap does not seem down to the questions on the IQ test being culturally biased against black people.

That leaves a range of possible explanations, none of which are comforting. James Watson, the co-discoverer of DNA, is among those who have claimed that genetic differences (nature) between the races can explain it. Other experts in intelligence point to the different environments (nurture) that kids of minority racial groups in the

US typically experience: profound inequality in socio-economic backgrounds and schooling, varying cultural expectations and more limited opportunity. As the study of identical twins shows, tough circumstances can drag down the cognitive performance of people whose genes should allow them to do better.

There is plenty of speculation on what causes the black–white IQ gap, but not much strong evidence to go on, and certainly not enough to be certain. So most neutral and objective researchers tend to sit on the fence. And commentators like Watson who leap off the fence and land firmly on one side show they aren't neutral and objective.

The controversy rumbles along and still flares from time to time, most prominently in a 1994 book called *The Bell Curve*, in which psychologist Richard Herrnstein and political scientist Charles Murray discussed racial differences in IQ scores and genetics, which was interpreted as arguing that little could be done to help those with the wrong genes to succeed.

It's not just mental differences between racial groups that are claimed. A small core of intelligence researchers dedicate their careers to trying to find differences in the average IQ between all sorts of people – from men and women and northern and southern Italians, to the Irish and everybody else. Some of these scientists remain obsessed by cranial capacity and have gone back and re-measured all those nineteenth-century skulls.

Most notorious was a psychologist called Jean Philippe Rushton, a Brit who made his academic career at the University of Western Ontario in Canada with a series of bizarre studies intended to prove intelligence was linked to race through size of genitalia. (When Rushton died in 2012, his own university boss described his work as 'not highly thought of'. Others labelled him a straightforward academic racist.)

Rushton (not a geneticist) was a strong believer the black–white IQ gap was down to genetics – about half at least, he reckoned – and he argued public policies to offer black children help in school were therefore a waste of time. But then Rushton, it seems fair to say, wasn't neutral. His politics were clear. For a decade before his death he ran the Pioneer Fund, an organization set up in 1937 to promote eugenics, the founders of which courted the Nazis, and which would later pay for political resistance to the American Civil Rights movement.

Given this landscape, it's easy to see why even the mention of research on the genetics of intelligence can make people uncomfortable. It certainly made the officials at the Centre for Talented Youth uncomfortable – so much that they discussed the request to access their records for so long that the project that wanted to use them finished before they decided what to do. They knew that a political mind-set is ready to use the results – any results – to confirm and fuel inherent prejudices. So it's important to say that, while IQ is largely a heritable trait, there is no serious evidence to support the claim that a racial difference in genes can explain the black–white IQ gap.

And despite fears that research on the genetics of intelligence will cement in the (flawed) biological explanations for racial differences in IQ scores, the early results support an opposite conclusion; even though intelligence is passed from parent to child, it is done in a way far too complex to break down in simple and significant terms that neatly contrast between groups of people. (We'll come back to this in a later chapter.)

Forget white people or north Italians or the upper classes, or whatever socially, ethnically or geographically constrained population someone wishes to favour. The most reliable group of people to sire intelligent children are, simply, intelligent adults (of all colours and nationalities), just as taller parents (of all colours and nationalities) tend to have taller kids. When it comes to intelligence, nature can be cruel and unsentimental, but it does not pick sides.

The legacy of all this bad science – and the pockets of biased research that continue – have poisoned the well for many psychologists who want to investigate the nature of intelligence and where it comes from. That helps to explain why many neuroenhancement techniques under investigation started life as treatments in medical projects, and even today are considered to be on the fringes of serious science. There is something uncomfortable about scientists who want to look at intelligence for its own sake. There is suspicion about their motives. People are reluctant to get involved or sometimes even to discuss it. The shadow of eugenics, even the likely mention of the word eugenics, or even human improvement, is enough to put them off.

But scientists have not always been so cautious. Before eugenics arrived to spoil everything there was widespread and relatively uncontroversial interest in how the brain could be modified to improve its functioning. Before eugenics, after all, there was electricity.

EIGHT

Current Thinking

The use of electricity to alter the functioning of the brain and body was widespread in the past. George Orwell, for one, might have been surprised how little we use it today. By the time he was shot in the throat during the Spanish Civil War in 1937, the medical treatment used to save and restore his voice included a routine blast of direct current, then known as electrotherapy.

While Thomas Edison's team in New York was building the first electric chair, medics at Guy's Hospital in London had established an Electrical Room to treat both physical and mental disorders. Electrical therapy was given to promote wound healing and to relieve pain and to try to treat various diseases including tuberculosis.

The ability to pass electricity through the skull and so into the brain offered the most potential to these Victorian pioneers. Nineteenth-century psychiatrists craved the respect they saw heaped onto their colleagues who specialized in physical ailments, and saw electrotherapy as the answer. Asylums of the insane gave them the opportunity to experiment, which they did with gusto. Patients with what we would now call depression, anxiety and schizophrenia sat with their bare feet in a bucket of salty water while electrodes were touched to their heads and spine.

Results were mixed, and the theoretical basis offered to support claimed clinical improvements was fuzzy. Some scientists said electricity acted as a fluid passed to the brain through the blood vessels. It was said to both increase and reduce blood flow; it was described as both a sedative and a stimulant.

Electrotherapy peaked when it was heavily used by both German and British scientists during the First World War to treat cases of the newly emerged shell shock and return men to the front line. It was generally given to the lower ranks and officers escaped. One British doctor who used it on soldiers argued it would be less effective on men of more senior rank, as their superior intelligence and education meant their mental conditions were more complex, and so they needed more sophisticated treatment.

In his semi-autobiographical novel, *Voyage au Bout de la Nuit*, which described his war experiences, the French writer and medic Louis-Ferdinand Céline described how an army doctor had 'installed a complicated assortment of gleaming electrical contraptions which periodically pumped us full of shocks'. The treatment, he wrote, 'had a tonic effect'. Céline was not the first to claim electricity applied to the head could go further than just treating and alleviating symptoms. Regular reports had surfaced by then about electrotherapy also enhancing mental performance.

The Dutch physician Jan Ingenhousz wrote that, following an electric shock in Vienna in 1783, he initially lost his memory and judgement. But, after several hours' sleep, he woke to find his 'mental faculties were at that time not only returned, but I felt the most lively joye [sic] in finding, as I thought at the time, my judgement infinitely more acute'. He said he could now see 'much clearer the difficulties of everything and what did formerly seem to me difficult to comprehend, was now become of an easy solution'. At around the same time, a German doctor treating a boy with electricity for malarial fever reported he became 'quicker of mind'.

In 1899, the French doctor Stéphane Leduc described how one of his patients, an elderly judge he had treated with electricity for facial paralysis, continued to request the treatment long after his symptoms improved. Electric current to the head, the judge claimed, improved his mental rigour.

The judge said he felt:

'. . . lighter and my ideas are more clear. I can concentrate my attention more closely upon my work. I struggle more successfully against the sleep-producing effects of long pleadings; I grasp more clearly the arguments which are advanced before me, and I can weigh them more exactly. In fact, I find my intelligence is brighter and my work is easier to do, and for that reason I come to you for an electrical

application whenever I am confronted by a fatiguing or
difficult piece of work.

Mainstream science rediscovered the technique in 1999.
Psychologists in Germany interested in finding new ways
to treat epilepsy used electrical brain stimulation to probe
working memory and motor learning. Their tinkering with
electric current and the brain was not popular with colleagues. 'It's fucking dangerous,' they were told. 'You should
stop this immediately.' And due to a shortage of volunteers
they were forced to experiment on themselves and their
families.

Since then, electricity has been applied to the brain to try
to change just about every cognitive function, with some
success. Probably the most well-known successful experiment was carried out by scientists in New Mexico. It gets
quoted a lot because it seemed to improve the way a group
of volunteers learned to spot potential threats by picking
out concealed objects. It gets quoted more because it was
paid for by the US military.

Before it deployed soldiers to Iraq, the US Army made
them play a video game called *DARWARS Ambush!* to simulate what they would encounter. The game got recruits to
scan virtual landscapes for potential threats – a sniper on a
rooftop or a bomb in an oil barrel – and taught them to do
it quicker and more accurately.

In the study, the scientists borrowed still images from
this virtual reality game and showed them to civilian volunteers, who were given a matter of seconds to scan them

for disguised or hidden dangers. They were told they were in charge of a mission, and the stakes were high. If they caused a false alarm by seeing a threat where there wasn't one then they were scolded for delaying the operation. If they missed a bomb, hidden sometimes inside a dead dog or a child's toy, then they were shown a simulated video of the explosion and its grim aftermath.

Most people struggled at first, but gradually learned and improved. And the scientists found electrical stimulation to the brain speeded this mental process. Volunteers who had a 2mA current applied to the right side of their skull, above their inferior frontal cortex or right parietal cortex, improved twice as fast as the others. (Although one dropped out because they said they experienced a burning pain.)

The effect lasted for at least an hour after the current was switched off, which suggests the stimulation might have provoked lasting change in the brains of the volunteers. As well as making neurons more responsive, such current is believed to increase the expression of proteins at the junctions between them. This might make them more prone to form connections, and see the brain moulded more easily into a lasting shape. The current, in other words, could make it easier for connections to form, and more likely that those connections would persist. Neurons that fire together, brain scientists say, wire together. Those connections, as we have seen, determine differences in cognitive performance and so intelligence.

* * *

There is lots of hype around the potential for electrical brain stimulation. Scientists hate hype, or at least they say they do. But they know a little bit of hype in a press article or radio show draws attention, and there is no such thing as bad attention. The only thing worse than being written about, for scientists who want grant money, is to work in a field never written about.

The way most scientists try to avoid hype is to include caveats, and to stress research is preliminary. But, equally, the universities and funders who pay most of these scientists' salaries want to know something might come out of the work they are investing in. So when scientists present their research – to bosses, politicians and journalists – they often engage in a kind of game in which they seek to emphasize the potential pay-off of their work, but avoid saying when those pay-offs could realistically be delivered.

The potential pay-off for research on brain stimulation is extraordinary. That's not hype. Here are some statements written about the possible applications of brain stimulation by proper scientists in professional academic books and journals, to be read by their equally proper peers:

> Contrary to the popular belief of 'no pain, no gain' [brain stimulation] has been shown to accelerate learning and skill acquisition in complex learning tasks that normally take a long time to master and in a range of fundamental human capacities from motor and sensorimotor skills to mathematical cognition, with minimal discomfort or adverse side effects.

And:

> Improved attention, perception, memory and other forms of cognition may lead to better performance at work, school and in other aspects of everyday life. It may also reduce the cost, duration and overall impact of illness.

And even:

> A future with people wearing portable devices helping them stay awake during nightshifts or while driving a car, or improving their motor coordination during an intense track and field training session is becoming a more and more plausible and socially accepted scenario.

We're not there yet, but the field is growing fast, and scientists are working to improve the techniques, with research to map brain function in more detail and equipment that promises more accurate current delivery.

Given the scope and boldness of the statements above, it's not surprising that people like Andrew, who we met at the beginning of this book, want to try brain stimulation for themselves. Largely outside bona fide research institutes and universities and beyond the reach of any regulation or control, Andrew and others like him are building brain altering equipment and using it on themselves. They swap stories, techniques and tips over specialist sites on the internet. They film their experiences and upload them to YouTube. They are attracting attention – the day after I met him, Andrew was due to be interviewed and filmed by CNN – and the underground brain stimulation movement is starting to poke its electrodes into the mainstream.

Until recently, someone who wanted to try DIY brain stimulation really did need to do-it-themselves. The kit – wires and battery chiefly – could be bought easily. But it tended to be a specialist pursuit. That changed in the summer of 2013 when a US company began to sell ready-made headsets. For £179, the company promised a quick and easy plug-'n'-play brain stimulator. And forget the noble goals, or not, of raising intelligence. Its target market was to speed the reaction times of computer gamers.

Officially, most scientists in mainstream research frown on the DIY community, and not just because many of the DIYers like to be known as brain hackers. Sombre articles in scientific journals warn of potential dangers to the inexpert crowd and scoff at some of the amazing effects many of the DIY individuals claim to have achieved. But it's a strange relationship. Brain stimulation is still a niche academic discipline, even within neuroscience, and the brain hackers are the scientists' biggest fans. They pore over academic papers and abstracts of talks to be presented at academic conferences, rifle through the details and search for experts who are looking into specific issues and conditions.

Even the mainstream scientists aren't fully sure what happened to my brain when Andrew fitted me with the brain stimulator and slid the switch to 'on' that day in his flat. But they think it goes something like this.

148

Electric current needs to flow in a circuit, hence the two electrodes attached to my head. One takes the juice from the batteries and floods it into my head, and the second soaks it up again and sends it back to the batteries. Part of the reason the electric chair is so messy and unpredictable is the human body isn't a reliable conduit for electricity. Bones, skin, muscles, hair – all of it puts up more or less resistance to the current, which ends up trying to find its own, quickest, way back out again.When Andrew directs the electric current from the first electrode onto the top of my head, the path of least resistance to the second electrode is through the narrow bridge of bone that arches over the top of my head. So most of the current heads across my scalp and never actually gets into the brain at all. Electrical brain stimulation, it turns out, is mostly electrical skull stimulation, and the skull doesn't do much in response. It warms a little, and the skin on the top gets a bit itchy.

In a grisly demonstration of this lack of electrical penetration, two scientists announced in 2016 that they had fitted an electrical brain stimulator to a human corpse. Like the US poet Walt Whitman, the deceased had left his remains to science, and this time science made sure it took advantage. Passing electrical current into the head of a cadaver in a laboratory is straight from the pages of Mary Shelley's *Frankenstein* but these scientists, if anything, showed the opposite effect: hardly any of the current got through into the brain, they said, certainly not enough to directly activate tissue in a living brain and make it work.

To detect the electricity coming in, the scientists placed

200 electrodes into the corpse's brain. But when they turned on the electrical stimulator, these brain electrodes barely noticed. Only about 10 per cent of the current applied through the electrodes pressed to the side of the dead head made it into the dead brain. To directly activate brain cells, the scientists estimated, brain stimulation would need to double the current applied to the brain from its standard of 2mA to 4mA. That's not recommended. One of the scientists tried 5mA of stimulation on himself and said the dizzying effect was alarming.

The study received a lot of attention, and was widely presented as showing electrical brain stimulation was a waste of time. But that's not true. For the goal of electrical brain stimulation is not to activate neurons directly, and nobody who does research with the technique ever thought it was. The effect is indirect: rather than making neurons fire, the extra applied electricity makes it *easier* for them *to be fired*. And that takes much less current – certainly the 10 per cent that made it through the corpse's head should be enough.

The current Andrew uses penetrates about an inch into my brain. That's not far enough to reach all useful regions, but it does cover a lot of the higher functions, which are controlled by my cortex – the wiggly furrowed layer on the outside.

Once inside my brain, the current needs to come out again. To do so, it struggles through the cells and blood of my grey matter and white tissue until it reaches the area underneath the second electrode that will carry it back to the battery and complete the circuit.

As the current continues to flow from the battery, it sets up a predictable dynamic inside my head. In a region of brain tissue under the first electrode current floods in. And in a distinct region under the second electrode, current pools to leave again.

Under the first electrode – the anode – the indirect impact on my neurons makes them more willing to activate. Exactly how this happens is unclear, but the current seems to nudge the neurons towards an electrical state known as depolarization. This makes them more sensitive to signals that arrive from other cells. So with the same amount of effort my brain can induce more activity in the charged region.

Underneath the second electrode – cathode – it's a different story. It's the polar opposite and the effect of the current on the neurons there is something called hyperpolarization, which makes the neurons less sensitive to incoming messages. This makes it *harder* for my brain to activate that region.

With correct and careful placement of the electrodes, that gives scientists the ability to turn bits of the brain up and turn other bits down. In his hands, Andrew holds a conductor's baton, which can quieten my brain's percussion and swell its brass section at the flick of a switch. So let's see what type of tune it can play.

NINE

The Man Who Learned to Cry

Attempts to stimulate the brain directly with electricity will always struggle with the skull. As the experiment with the corpse demonstrated, most of the current from electrodes pressed to the outside of the head doesn't pass through into the brain. Up the amount of current and the itchy tickling on the scalp worsens to an unpleasant burning sensation. Up it further and the burning is not just a sensation.

For better penetration, some neuroscientists have used lasers. Researchers at the University of Texas took a low-level CG-5000 medical laser, approved to improve circulation and relieve muscle pain, and pointed it instead at people's foreheads. They were trying to activate an enzyme in the frontal cortex called cytochrome oxidase, to help brain cells produce more energy and so work harder. It seemed to work: volunteers given the laser treatment performed better on tests of memory and attention.

Another answer is to use magnets. Michael Faraday famously realized that waving bits of metal around inside a magnetic field can induce electric current and he used the discovery to invent the electric motor. Faraday's lectures and demonstrations at the Royal Institution in London drew such large crowds that the road outside – Albermarle Street – was made the first one-way street in the city in an effort to control the traffic.

Faraday's work with magnets also brought him to the attention of a different crowd; the followers of a controversial doctor called Franz Anton Mesmer. Unlike Faraday, we don't remember Mesmer for his brilliant science. That's because he didn't do any. But he was still influential. He left us the terms 'mesmerize' and 'animal magnetism'. And he developed a link between them, which we now call hypnosis.

Mesmer claimed human disease was caused by the movements of the sun and moon, which disturbed tides of invisible fluids in the atmosphere and inside the human body. The nervous fluid inside people was magnetic, and the imbalance in this animal magnetism caused by the motions of the heavenly bodies, Mesmer said, could be fixed by applying magnets to the body of the patient.

Mesmer was the first stage hypnotist. Demand was so great for his magnetic therapy he would treat dozens of people at a time, tying them together and making them stand around a specially constructed apparatus called a baquet. This was a circular wooden case, about a foot high, which Mesmer and his helpers, chosen for their 'youth and comeliness', heaved into the centre of a large hall and filled with powdered glass, iron filings and symmetrically arranged bottles, usually all covered in water. From holes in the wooden lid, long iron poles stuck out.

Mesmer's patients, up to thirty at a time, would sit in a circle around the baquet, holding both the poles and each other's hands, while Mesmer walked among them with an iron rod. The hall was hung with thick curtains, and the silence Mesmer insisted upon was broken only by gentle music played on a pianoforte or harmonicon, accompanied sometimes by singing.

Wearing a lilac coat, Mesmer would strut and sit beside his patients for two to three hours, fixing their gaze on his, and prodding and stroking their diseased bodies with his iron stick. He would place his hands on their stomachs, or, shaping his fingers into a pyramid, move from their heads down to the feet and back again.

The patients? Some barely reacted, staying calm and claiming to experience nothing. Others coughed, spat, reported slight pain, a local or general heat, and fell into sweats. Some went into convulsions called crises. Mesmer's theatrics seem to have had an unusually strong influence on many of his younger female patients.

Among those influenced by Mesmer and his ideas was IQ-test pioneer Alfred Binet, who dabbled with magnetism in his early years. After he witnessed one display, Binet wrote: 'Young women were so much gratified by the crisis they begged to be thrown into it anew; they followed Mesmer through the hall and confessed it was impossible not to be warmly attached to the magnetizer's person.'

As other magnetizers copied Mesmer's techniques, another strange consequence emerged: some patients went into a passive, trance-like state, during which they appeared asleep but could nonetheless continue to listen and talk to the hypnotist. In this 'magnetic sleep' the patients became prone to suggestions – women could be made to fondle and kiss an imaginary baby and men to pretend to be drunk. It was nothing to do with the magnets of course and stage hypnotists have repeated the trick ever since.

It was an inauspicious start for magnetic stimulation, but other scientists were drawn to it and continued to experiment with the way it could influence the body and the brain. And just like the use of electricity, magnetic stimulation of the brain has enjoyed a scientific revival in recent years.

In 2008, for instance, a man with autism called John Elder Robison had fluctuating electromagnets applied to his head. As Faraday predicted, the combination of the magnetic field and the movement induced electric currents inside John's brain – much stronger current than could be achieved by direct electrical stimulation. As a consequence, something in John's brain was released. There were no lilac coats this time, but the effect was mesmerizing.

John was taking part in a research study at Beth Israel Deaconess Medical Centre in Boston into how people with autism process language. He had thirty minutes of transcranial magnetic stimulation (TMS). Because the focus of their research was language, the researchers targeted Broca's area, part of the frontal lobe. The scientists asked him to wear a gumshield during the magnetic stimulation, in case of involuntary movements. They told him any effect would probably be mild and short-lived. They were wrong.

The first sign something had changed for John was during a telephone conversation later in the day. His voice sounded different to him. He was using different tones and was lowering and raising the pitch at the end of words for emphasis. He realized with astonishment he was doing so to portray emotion. Like many people with autism, John had previously struggled to decode and identify emotion, and to appreciate how the tone of someone's voice could say as much as their choice of words.

Now, for the first time, John's voice sounded to him like it carried an emotional range. Confused (though probably not as confused as the friend he was talking with), John hung up the phone. He put on some music – an old track by the Tavares, a group John had worked with in a previous career as a sound engineer.

'Everything was different. Every little nuance of the recording held meaning for me. My range of sonic comprehension had just widened a thousand-fold. Whatever they did with that brain stimulator had unlocked something very powerful in the way I heard music.' The magnet seemed to have released something inside his brain and the effect went beyond music.

'The filter of autistic disability – if that is what hid the emotion from me before – seemed to have vanished. I heard a smile in one voice, as I saw it on my friend's face, and I felt its truth inside of me.'

The scientists who had delivered the brain stimulation were as surprised as John at his transition.

A few hours – all the time that had passed since the TMS

session – was nowhere near long enough for John's brain to have plotted and formed the new connections to allow him to experience emotion in this sophisticated way. The capacity must have been in there all along. The brain stimulation, somehow, had released it, had switched it on. They asked John to tell them if anything else unusual happened. It did.

Faces, other people's faces – previously an inscrutable mask – started to talk to him. John was working as a mechanic and he realized that a female customer was communicating with him, but in a way he had never tuned into before.

'As she spoke, her face began to tell its own story. I wasn't even hearing her words, but her feelings shone through clearly.' As the woman's words spoke of the mechanical problems with her car, John could read a deeper narrative in her expressions and tone. She was anxious, about the car, the cost of the repair, her ability to pay, how she was going to get to work.

Like many people with autism, John had spent a lifetime blind to these social cues – the kind the rest of us take for granted. Previously, he would have answered her with a non-committal factual response that she would simply have to drop the car off and wait. Instead, from out of nowhere, the newly engaged part of his brain blew away the dust gathered over decades of inaction and answered for him.

'Don't worry. The problem you are describing sounds like a pretty small thing to fix.'

Not everybody with autism wants a fix, but plenty do.

When John wrote about his experience with TMS on his blog, lots of them got in touch. Some wanted advice, some wanted to try the brain stimulation for themselves, or try it on their children. All wanted some hope.

John's unlocked emotional frequencies did not all bring good news. He found it difficult to turn off his new awareness, and he would burst into tears when he read in the newspaper about the deaths of total strangers. A fire hose flow of unregulated and unfamiliar emotion and empathy now flooded his consciousness. 'Experiencing the collective emotional energy of a small crowd and feeling each person's hope, fear, excitement and worry was just as disabling as being blind to it.'

The brain stimulation had given John a new ability. He felt like he could see into people's souls. He was desperate to know more about what had been done to him, but the scientists involved could only speculate. The magnetic stimulation had been low frequency, which is known to inhibit brain activity in the same way as the cathode does in the circuit set up during electrical stimulation. (High frequency magnetic stimulation has the opposite effect and activates brain cells.)

Perhaps, they suggested, John's brain had reacted to the inhibition by bringing other circuits online to compensate for the temporary loss. Maybe the surge of activity through a long-idle part of John's brain had jump-started a dormant mental ability to sense and judge emotion.

It's easy to be sceptical of the changes John reported. We have no independent way to confirm them beyond John's personal testimony, and in most branches of science and medicine, such anecdotes tend to be scoffed at unless they can be backed with convincing data and the effect preferably repeated in lots of other people.

But remember one thing: the entire field of psychiatry and the study of mental disorders pretty much rely on personal testimony. Millions of people are diagnosed and treated for depression and OCD and anxiety and dozens of other problems depending on the answer they give to the question: so how do you feel? There are no brain scans, blood tests or physical measurements to probe the state of our minds.

John's discovery of emotion, and the unlocking of his ability to read people, certainly makes a good story. He has written a book about it, called *Switched On: A Memoir of Brain Change and Emotional Awakening*, the latest in a series he has published on his life and experiences with autism. So, might John be, well, exaggerating? Could the change be not in his perception of emotion but in his imagination?

Only he knows for sure, but it seems unlikely to me that he is not telling the truth. For one thing, John wasn't the only person with autism to report these effects after the TMS sessions. A woman called Kim who had enrolled in the same scientific study contacted him to share her identical experience. She too found for the first time she was able to read and judge people's expressions. She knew when people were speaking in a sarcastic way. Social interaction, previously a black-and-white affair, was now presented to her in glorious Technicolor.

She wrote on John's blog: 'Before this stimulation, I thought that I read people's facial expressions and their tone of voice fairly well. However, after seeing the difference following the stimulation, I would say that I miss 50 per cent or more of the social interaction.'

The plural of anecdote, as the online debunkers of pseudoscience like to point out, is not data. Kim and John did not report their experiences independently. It's possible they influenced each other, and gave each other more confidence to over-interpret the change. Still, if they were fooling themselves, then they did so in a bizarre and pointlessly self-destructive way. Both John and Kim said they were changed by the stimulation, but not all for the better. Kim was distraught when she revisited her memories with her new emotional range and realized what she had missed out on.

'Suddenly I understood why I have trouble with my friends, and why I don't get along with my co-workers. TMS showed me everything I had done wrong in my life and it overwhelmed me.'

Things got worse when her new ability quickly faded away, her multicolour world snapped back to monochrome.

'What am I going to do now? It's like I'm haunted. I got a glimpse of those emotions but now it's gone. So now I know what life is like for other people but it's not that way for me.'

John also revisited past experiences and reassessed

relationships and, as he did so, realized one of his friends was not what he seemed. What John had always thought was friendly chit-chat with him he now identified as ridicule and belittlement intended to single him out as different. John pledged never to speak to him again. And more followed: the TMS changed John's mind in such a fundamental way it broke up his marriage.

Martha, his wife, suffered from serious depression. Sometimes she felt so bad she struggled to get out of bed. For years, her illness had not been an issue for John. He simply left her to it. He had shared her life, but not her sadness, because he didn't do other people's sadness.

The brain stimulation made him capable of seeing and feeling his wife as she truly was. And to his shame and distress, he realized he could not handle the emotions they now shared. The cloud of misery she lived in started to engulf him too. When he was with her, Martha's depression would crush him and he felt dragged down. The feeling only lifted when he went out and left her in the house. And one day he never went back.

'TMS took away my emotional innocence, and I'll always be sad for its loss. But part of the cost of getting smarter emotionally was seeing people as they actually are, and not as I imagined them to be.'

John lives in Massachusetts and I spoke with him on the phone just before Christmas 2015. He's articulate and easy to talk to and offered some thoughtful reflections on his experiences. He was also – and in my experience as a journalist this is a sign when someone is on top of their subject and telling it like it is – honest enough to admit there were some things he couldn't explain and he didn't have the answers for.

The change is still there, he said. It's not as vivid as it was immediately after the stimulation and it sometimes needs to be nudged into action – he can miss some social cues unless he is primed to watch for them – but John is convinced, and convincing: the magnetic stimulation of his brain brought fundamental change.

Here's an interesting question: if we accept that John experienced the change he describes, then did the magnetic stimulation make him more intelligent? After all, he can judge and respond better to environmental cues based on emotions now, so he is in a better position to use what he has got to get more of what he wants. But better reading of emotion probably wouldn't help him on an IQ test. So does intelligence go further than IQ? How far? Can we use neuroenhancement, therefore, to improve 'intelligence' even if it has no impact on IQ? The answers, just like most in this fascinating and complex area of science, are far from simple and extremely controversial.

TEN

The Brain and Other Muscles

While g, the general intelligence capacity identified by Charles Spearman, is critical, it is not the only psychological factor that goes into determining your intelligence. Indeed, Spearman never intended it to be viewed as such. He left open the possibility for specific mental skills to vary from task to task. Two people with the same high g, he reasoned, could still show variable performance in, say, music and French.

It's not enough to simply have the intelligence, we have to apply it. And, naturally, some people are better at applying it to some tasks than others. Spearman, who liked to keep things alphabetically simple, called this extra variable 's' for specific intelligence.

In this model, g is the raw power, the size of the engine. But s measures how well this power can be channelled into each action. A four-wheeled Ferrari is impressive on the road but in the sea? Not so much. In that (admittedly crude) example, the Ferrari has a high g and a high s for driving, but a low s for swimming. It can't use its high-performance engine that way. It's still a sleek, beautiful and powerful car. It still has a high g in the water. But it sinks.

The introduction of specific s factors means intelligence emerges from a hierarchy of diverse mental abilities. This is important for the idea of cognitive enhancement, and how it could be achieved, because it suggests more than one way to increase someone's ability. The first way, in theory, would be to target and increase the overarching g. As g is based on a natural capacity, that seems a pretty large challenge. The second possibility is to intervene to improve one or more of the s-factors – to change how the brain accesses and uses that capacity. And that seems more do-able.

Probably the most well-known architecture of intelligence splits the influence of g into two measures of cognitive ability: crystallized intelligence and fluid intelligence. Crystallized intelligence, as its name suggests, is the crunchy stuff deposited in our heads over years. It's the knowledge, the dates and lists of kings and queens. Did you know the capital city of Cameroon is Yaoundé? If not, then I have just slightly increased your crystallized intelligence, provided you remember it of course. Memory and recall are important for crystallized intelligence, and understanding and manipulating numbers too. Most of all, crystallized intelligence is vocabulary – having, using and making sense of words.

Fluid intelligence, also well named, is the cognitive processing we tap to solve a problem. It's the ability to reason, to make connections, and to make use of the crystallized knowledge. It's the detective work – analysing the clues and making deductions.

Just like with the exam scores that pushed Charles Spearman towards the discovery of g, levels of crystallized and fluid intelligence tend to correlate tightly. It's unusual to find a person with very high levels of one and very low levels of the other.

Some psychologists break g down into an additional third output: spatial awareness, including navigation and the ability to hold and manipulate visual imagery in the mind. This kind of intelligence is more common in men than women. Women get their own back by having better short-term memories.*

Not all scientists accept the idea of g as dominant general intelligence. The most high-profile challenge came from the psychologist Howard Gardner in the 1980s. He took the specific intelligence idea to its logical extreme and argued specialist abilities were the driving influence on cognitive performance. Indeed, he said, s was so important that the effect of g was small and need not exist at all. S alone mattered and, without general intelligence to tie types of specific intelligence together, each of these multiple intelligences could be individually high or low. More fundamentally, Gardner claimed each specialism was a different type of intelligence. There were multiple types of intelligence a person should be assessed for, he said, not just one.

---

* On average, of course.

Some of these different types of intelligences that Gardner outlined look similar to what we think of as aspects of Spearman's general intelligence. Two, for example, are called logical-mathematical intelligence and visual-spatial intelligence – these reflect what seem to be standard cognitive skills, and the kind already measured by IQ tests.

But Gardner also introduced less conventional types of intelligence, from musical intelligence, interpersonal intelligence and naturalistic intelligence to bodily-kinaesthetic intelligence and mental searchlight intelligence.

What are these? Bodily-kinaesthetic intelligence is attributed to people who are especially skilled at using their body to convey ideas and feelings. They are aware of their presence within physical space, rely heavily on their sense of touch and have good motor skills and hand-eye coordination. Dancers and athletes, the theory says, show high levels of this type of intelligence. Gardner's mental searchlight intelligence is the ability to scan lots of sources of information at once, to make sure nothing is missed or missing. Naturalistic intelligence is sensitivity to the animal and plant kingdoms, such as shown by gardeners and zookeepers. People who are clever in an interpersonal way are sociable and good mixers and enjoy helping others.

It's a compelling idea and in many ways the theory of multiple intelligences paints a reassuring portrait of humanity. We all have something we are good at. We're all equal. Teachers and educators love the idea of multiple intelligences because it makes every child clever in their own way. It offers a comforting view of the world, similar to seeing the chaos of romantic love and relationships through the rose-tinted lens of 'someone out there for everybody'. This appeal has helped make the idea of multiple intelligences well known. But as scientific theories go it's controversial and pretty flimsy.

Its social and political appeal rests on how it spreads performance and ability (and, tacitly, value) around. But for this to be true, then each of its various types of intelligence should be truly independent of each other. People who are good at logic puzzles should not be any better at spotting patterns, say, than someone who is not. People who are skilled at musical instruments should have no advantage when it comes to spatial awareness.

Most studies suggest the opposite: good and bad performance on separate tests of these so-called multiple intelligences tends to bunch together, in the same way Spearman found academic grades did more than a century ago. The same people still tend to do well on most of the tests of Gardner's different types of intelligence, and the same people tend to perform poorly. Despite the theoretical attempt to pull the skills and abilities apart and spread the results across the population, the data puts them back into sticky clumps and hands them, fairly or not, more to some individuals than others.

Still, the popularity of the idea of multiple intelligences – all shall have prizes! – has spawned a series of imitators, most of which, in scientific terms, are little more than fashionable labels. Entrepreneurs write and sell books on business intelligence and managerial intelligence. There is spiritual intelligence and existential intelligence and moral intelligence and sexual intelligence and leadership intelligence. There is people intelligence and cultural intelligence and narrative intelligence and creative intelligence. There is even a dark intelligence, made up of an unholy trinity of personality traits: narcissism, Machiavellianism and psychopathy.

One thing many of these claimed types of intelligences have in common is the way they are presented as an alternative to 'conventional' intelligence, as measured by IQ tests. They are sold as more reliable indicators of human ability and potential, or at least a more useful guide to how someone will succeed in work, relationships and society. That is especially said to be true for emotional intelligence. We hear a lot about emotional intelligence, usually in the negative – 'oh, yes, he's *academically* bright, but he isn't *emotionally* clever'.

Emotional intelligence is a real thing and it has a legitimate scientific foundation. But as part of the push-back against what is seen by critics as the tyranny of IQ tests, emotional intelligence has also been twisted and turned into a conceptual woolly blanket used to comfort people who believe they can't do mathematics.

Probably the greatest reason why emotional intelligence is so well known is the 1995 book of the same name by the psychologist and journalist Daniel Goleman. The subtitle is, 'Why it can matter more than IQ'. The blurb on the back says the book 'redefines intelligence'.

Goleman's book highlights emotional intelligence and

other rival intelligences as important abilities (which they are), with a role in human performance (which they could well have). But it also goes further and explicitly positions them as *superior* measures of mental and cognitive abilities, different from and more important than IQ (which they're not).

This attitude is common and it feeds on all those fears of IQ, typically presented as an elitist establishment idea and a private members' club that turns away people at the door. Rival intelligences, their inventors claim, are more inclusive, more open and – crucially – more malleable and changeable. For what use are ideas like business intelligence and sexual intelligence if they cannot be increased in exchange for the price of a book, DVD or conference ticket?

Rivals to IQ also trade on the idea they are more relevant, they measure separate and different abilities, which, although they are called intelligences, are more useful to have than 'intelligence'. They are presented as independent of 'academic' intelligence – it doesn't matter how you did in tests and exams at school or if a teacher or friend was once rude about your brain power, you can still make something of yourself.

That's true and admirable, and of course if people can learn to improve their business, managerial, creative, sexual, people, narrative, cultural, spiritual, moral and existential skills and awareness, then they are more likely to do well, to achieve their goals. But it's misleading to present these opportunities and abilities as distinct from IQ, or general intelligence, even more so to present them as rival forms of intelligence.

When Howard Gardner first introduced his multiple intelligences, he admitted he used the word 'intelligences' rather than skills or abilities because it would draw more attention. In Daniel Goleman's book, he wrote that:

> There are widespread exceptions to the rule that IQ predicts success – many (or more) exceptions than cases that fit the rule. At best, IQ contributes about 20 per cent to the factors that determine life success which leaves 80 per cent to other forces . . . My concern is with a key set of these 'other characteristics', *emotional intelligence* . . . No one can yet say exactly how much of the variability from person to person in life's course it accounts for. But what data exists suggest that it can be as powerful, and at times more powerful, than IQ.

That's not true. As we've seen, the data, where it exists, shows a strong link between IQ and a person's life course, at least in terms of their achievements. The stranglehold of Spearman's positive manifold on mental ability means emotional intelligence, and indeed any kind of intelligence that truly flexes the brain, must link to most of the other kinds, including those measured by IQ tests. Measure one kind of intelligence and you get a pretty good guide to how well people perform on others.

Take bodily-kinaesthetic intelligence. It's about as far as one can get from the pencil-and-paper impression of IQ.

But the correlation with scores on standard 'academic' measures of intelligence is still there. Tests show how well someone can mentally control how they move their arms and legs, and how they judge speed and movement, and even how they kick a ball, can also indicate their broader mental skills. Say hello to the intelligent footballer.

When I used to watch a lot of the sport in the 1990s, there weren't any intelligent footballers. Well, there must have been, but they didn't tend to make themselves known. It's not hard to see why. Poor old Graeme Le Saux, the one-time Blackburn Rovers, Southampton, Chelsea and England defender, was targeted from the terraces for being a bit of a clever clogs (and for reasons that always escaped me, homosexual) simply because he had a couple of A-levels and read the *Guardian*.

As television money and foreign talent swarmed into football, so the game filled with sophisticated and urbane continentals who could speak several languages and – shock – ate pasta. 'He's got a good football brain' was added to the list of commentator-friendly attributes affixed to these foreign types who could lift their head up and pick a pass under pressure rather than booting the ball into the crowd.

Tactics and roles evolved. While former players asked to analyse matches had once been able to speak in that weird mix of past and present tense only footballers seem to use – 'I've seen him running and he's crossed it in and I just hit it' – today they are expected to demonstrate knowledge and insight. The bar has been raised so far that in 2013, the former top-flight attacking midfielder Paul McVeigh published a book titled *The Stupid Footballer is Dead: Insights into the Mind of a Professional Footballer*. Cynics of the subtitle please note: the book extends for 160 pages.

It might not be the same as sitting an IQ test, but playing high-level football – or any team sport – demands plenty of cognitive ability. Each player needs to observe, think and react quickly, and accurately make and test mental plans. Sports psychologists use terms like visual anticipation, knowledge of situational probabilities and strategic decision making to describe these skills, which can sound specialist and relevant only to their sport. But as we've seen, mental ability doesn't usually work that way – people who are good in one regime are usually pretty useful in others.

Certainly, there are plenty of outstanding all-round sportsmen and women, which itself shows impressive cognitive flexibility. There are footballers who excel at cricket and golf, and racing drivers who are expert skiers. (Few took it as far as Max Woosnam, the English sporting polymath who won a doubles tennis title at Wimbledon, scored a maximum 147 break in snooker, made a century at Lord's Cricket Ground and captained Manchester City football club.)

With a little imagination, terms and language of sports psychologists can translate to describe mainstream mental skills that apply in the wider world: spatial attention, divided attention, working memory and mental capacity – all combined with the ability to change strategy and inhibit responses. Another way of describing this group of cognitive tasks is executive function. And good executive function is useful way beyond the sports field.

In the summer of 2007, scientists in Sweden recruited dozens of players from the country's elite football leagues to test their intelligence. Coaches at several clubs, from the top and a lower division in the men's and women's sport, were asked to nominate two of their defenders, two midfielders and two strikers to spend forty minutes sitting a series of mental tests. They weren't IQ tests – the tasks didn't analyse language skills – but the questions were standard psychological measures of executive function. In one puzzle, called Design Fluency, the footballers were given sixty seconds to find as many different ways as they could to join all of the dots in a square with a single continuous line.

The tests were anonymous, so we don't know which footballers were volunteered by their coaches (perhaps those who played poorly in the previous game?). But to give an idea of the calibre of player involved, those who started games in the Swedish top division that season included Henrik Larsson, the former Celtic and Barcelona striker who played in three World Cups, and Stefan Thordason, who scored one of the best goals I have witnessed live, for Stoke City in a cup match at Charlton Athletic.

The results of the study were clear: all the footballers did better on the mental tests than the average person would, and the top division players scored in the top 5 per cent of the population. What's more, the performance on the cognitive tests seemed to predict future success on the pitch. The smartest players scored or helped to create the most goals in subsequent seasons.

The scientists were so struck by the results they suggested football coaches might be missing a trick by focusing only on physical ability and technical skill when they assess and recruit young players. A quick thirty minutes of pencil and paper tests, they suggested, alongside the shuttle runs and free-kick expertise, could be a useful way to predict which youth players will make the grade. The stupid footballer may not yet be dead, but he's being pushed aside by more intelligent team mates.

There is an opportunity here. If the workings of the brain can influence sporting performance, and the workings of the brain can be improved with neuroenhancement, then smart drugs and brain stimulation should be able to help athletes to compete by upping their intelligence. Just like the DIY brain hacking community, plenty in the field of sport are trying it. Physical doping in sport has been joined by brain doping.

The cyclist Tom Simpson died on the Tour de France because he turned off his basic survival mechanisms. The drugs in his system changed the way his central nervous system responded to the physical exertion, to the demands on his physiology. This allowed him to push his body's performance beyond what the brain would usually allow, with, as it tragically turned out, good reason. Neuroscientists are trying to use their new tools of brain intervention to achieve the same result, but in a safer, more controlled way.

Stories surface from time to time of what is called miracle-strength – mothers who lift cars to save their trapped children and so on. We should be sceptical. Those reports remain unconfirmed and, by their extreme and unusual nature, untested. A definite physical limit restricts what a human body is capable of, whatever the circumstances. But there is also a mental limit. And often the mental limit is set at a lower threshold than the physical limit. To protect us from danger, the brain tells us we are tired before we are. It does this by signalling we are exhausted, that we have reached our physical limit before we have. How else can the winning athlete, who has given everything to cross the finish line, then set off on a sprightly lap of honour?

Sports scientists call this the central governor theory. The central governor likes to play it safe. When the brain senses the body approaching potentially dangerous levels of exertion – heart rate, blood pressure, oxygen demand, muscle fatigue – it sounds the alarm and convinces us we are simply too knackered to continue.

Much of sports psychology and training aims to exploit the zone of what is physically possible even after the central governor tells you it's not. It's the pushing through the pain barrier, silencing negative thoughts, getting in a positive mind-set. Doing so is usually presented as a question of motivation, from the Olympic swimmers who listen to music on chunky headphones even as they approach the blocks to my friend who, training for his first marathon, said the hardest part of the long lonely runs in the months before was resisting the voice in his head that said, 'Look, there's a bench. Why not sit down?'

In theory, neuroenhancement offers a way to silence this voice, or at least turn it down. By directly interfering with the way the brain works, the threshold of the central governor could be increased, or the muscles could be told to work beyond it. And so electrical brain stimulation could offer a way to push physical abilities and enhance the mental side of athletic performance.

There's some evidence for this. In 2013, scientists in Brazil found twenty minutes of electrical brain stimulation of the brain's motor cortex, which controls muscle movement, increased the performance of trained road cyclists on something called the maximal incremental exercise test. It's the athletic equivalent of testing to destruction. Each cyclist was placed on a static bike, and, as they pedalled, the resistance level was increased every minute. The test finished when the cyclist 'voluntarily terminated' the exercise, or because they couldn't keep up with the required speed of spinning the pedals at 80 revolutions per minute (rpm).

The highest intensity each cyclist could sustain for a full minute before they stopped – voluntarily or not – is called the peak power output. The trial worked: motor cortex stimulation increased peak power output by 4 per cent. That doesn't sound much but, just like small increases in intelligence, it could be the difference between success and failure in a competitive race.

The technique might help the less committed too. In 2015, another Brazilian experiment tested the same effect on men in their twenties who were merely 'physically active' – defined as taking some exercise three times a week. This experiment had a more fearsome name, the time to exhaustion test, and several of the participants never made it as far as the bike. Four of the initial fifteen volunteers dropped out, at least one because he was scared of having his brain Westinghoused. The survivors were given brain stimulation to the motor cortex again (or not) and simply asked to pedal at more than 60rpm at a fixed (pretty tough) level of resistance. When they fell below target speed for five seconds, they were labelled as exhausted.

Without the brain stimulation, the weekend cyclists managed an average of 407 seconds. After the electric current to their brain, they could keep going for more than a minute longer, and on average lasted 491 seconds.

I still had a few months before I was due to return to Mensa, so I thought I would give it a go – to see if I could stimulate my brain to improve my physical performance, before I tried to boost my mental performance. To do so, I bought my own electrical brain stimulator. The device marketed to computer gamers was sold out, but a quick internet search threw up a number of other companies selling their own ready-made versions. I went for the cheapest. It cost $55 and was posted to me from America within a fortnight.

Setting up was easy. A 9V battery, one of the chunky rectangular ones, fitted snugly inside a white box with a socket on the outside to receive the wire to connect the two electrodes. Each electrode was colour coded, red for the anode and black for cathode, and each ended in a crocodile clip, to be attached to a saline-soaked sponge that would transfer the current to the outside of my head. The switch on the box had three settings: off and then a choice of 1mA or 2mA of current. That's about enough to light the small standby bulb on your television.

If you take it seriously enough, you can spend lots more – both on the stimulator itself and on bespoke sponge electrodes and ready-made saline solution to soak them in. Professional versions – the types used by research scientists – cost more still. They promise more reliable and controllable current and more accurate electrode positioning, but they all work on the same simple principle.

My budget version came with a couple of pages of instructions that said to chop up a regular household sponge and wet it by mixing a couple of spoonfuls of salt into a cup of water. Placing the soaked sponges in the right place next to my skull and keeping them there was tricky (I had spurned the chance to purchase the special headband) so I rummaged in a drawer until I found some close-fitting headgear. It was a knitted Spiderman hat. Other hats are available.

Where to position the electrode sponges? Unhelpfully, the instructions that came with the equipment said company policy was not to recommend any electrode positions. 'A quick Google search', they promised, would provide guidance. Although, the instructions also pointed out, the 'content or validity' of these websites could not be guaranteed. This was true DIY brain stimulation – users are advised to 'research and come up with your own conclusions on how you will use' the device. It was, the instructions said in bold for added emphasis, '**in no way a professional medical device**'. Avoiding any claims for benefit has, so far, allowed the manufacturers of brain stimulators to avoid regulation.

A warning: should you wish to try brain stimulation for yourself, there are a colossal number of academic studies on brain stimulation that a 'quick Google search' throws up. For at least a couple of decades, plenty of neuroscientists have made a career out of scanning the brains of people while they are asked to read or say something, to think of words and pictures, to taste drinks and even while they are sexually stimulated. In this way, neuroscientists have mapped parts of the brain they say are associated with just about every human cognitive function.

A new generation of neuroscientists is now going further and using these maps to investigate brain stimulation. Regions called the dorsolateral prefrontal cortex and the temporo-parietal junction, for example, have been shown in brain scans to be involved in the way we form moral judgements. So, naturally, scientists have tried to stimulate these brain regions to see if it changes how people make these judgements. The left frontal region is known to be involved with language formation, so scientists have tried to stimulate it to see if it helps people say the tongue twister 'if two witches would watch two watches, which witch would watch which watch?'

Brain scanners are expensive and usually housed in major universities and research centres. That doesn't guarantee the quality of the research done with them, but it does sometimes help vouch for the credentials of those who carry out the studies. However, as my experience shows, any fool can buy and experiment with their own brain stimulator.

There are plenty of robust and careful brain stimulation experiments accurately written up out there in scientific journals. And there are plenty of misleading studies that have limited statistical power, or are fundamentally flawed. Unfortunately, neuroscientists have yet to identify the part of the brain that allows non-experts with a 'quick Google search' to tell the difference.

The scientists who did the endurance bike tests used a Veletron Dynafit ProTM cycle simulator. I don't have a Veletron Dynafit ProTM cycle simulator, but I do have a Concept 2 rowing machine, which promises 'the ultimate all-body work out'. A gym I once visited in Cardiff had a sign on the wall: 'Rowers exercise. The rest just play games.'

I bought the rowing machine when I had kids and realized I would be spending more time than before in the house, and I use it pretty regularly. You pull against a flywheel, so it uses your own effort against you. It's noisy, but that's not the biggest annoyance with my rowing machine. The biggest annoyance is the digital display of your performance. You can set it up with a pace boat to race against, but the cold hard numbers are usually enough. Any sub-conscious, involuntary slowing in the push of the legs or the tug of the arms, and the display screen reacts instantly to show you are weakening, sometimes before the desire to slow down has even registered. And it *hurts*. When I started to read up online about training sessions and stuff, I saw lots of references to breaking the seven minute barrier, to cover a distance of 2000m. There are web discussions, hundreds of pages long, dedicated to the milestone and how to achieve it.

Without getting too bogged down in the details, it's probably enough to say the only way to achieve it is row like the clappers for the first 500 metres, and then try to keep going, ignoring the physical and mental signals that flood your muscles and brain and tell you for the next 1000 metres you are going to die if you do. At 1500 metres, death loses its sting and the digital countdown of the distance remaining becomes the centre of the universe. With 200 metres to go, about thirty-five seconds' worth, the universe explodes. Eyes bulging, nose streaming, brain-dissolving agony remains, and a thought bounces around your head and swells to a crescendo: If I Don't Stop Then I Never Have To Do This Again.

I managed to row 2000m in under seven minutes, just, a few years ago and I have kept the promise I made to myself in that last 200 metres, and never done it again. I wasn't going to do it now, even with my brain stimulator to help. Instead, I set up a test to row as far as possible in four min-utes. It's a test to exhaustion, or at least it is when I do it.

I would do the four minute trial twice, once with the electric current massaging my motor cortex, and once without. To try to make the comparison a fair one, I asked my wife to help me – to decide on which trial the stimulator would be turned on. I would wear it both times, so I wouldn't know. I also covered up the screen with some black tape, so I could only see the reducing time. Knowing the distance, I figured, might skew the results by giving me a target to aim for on the second run.

I dutifully dipped my sponge electrodes in my home made saline and found the Spiderman hat. My wife fiddled with the switch and I gave her the four minute warning and rowed as if my life depended on it. A couple of hours and a couple of bananas later, I repeated the test, again asking my wife to turn the stimulator on or off. The second time felt easier if anything, so I was surprised when I unpeeled the tape and saw the results. They were pretty much identical – 1,152 metres on the first and 1,148 on the second. Enough, according to the online charts of performance, to comfortably make the 'above average' category but not enough to be rated as 'good'. I could live with that.

I looked at the stimulator switch; it was set to 2mA. So, the second test had been the one with the help.

'It didn't make any difference,' I said. 'The electric current. It didn't make me row any further.'

'Well, how do you know?'

'The results, they're almost the same. It was switched on during the second test and I didn't do any more.'

'But what about the first one?'

'Well, it was turned off, right?'

'No.'

'What?'

'You asked me to choose so I turned it on both times.'

As breakdowns in scientific communication go, this wasn't up there with the you-use-metric-measurements-and-we'll-use-imperial-units disaster that saw NASA's $125m Mars Climate Orbiter fly into the red planet in 2000, rather than around it. But it did mean my effort was wasted. And I didn't feel like doing it all again.

I decided on a different approach. I would do what good scientists do and try to prove my hypothesis wrong. The idea – brain stimulation could help me row further – could now be easily disproven. I had a target to aim for. If I ripped away the masking tape and the electrodes and went for it unaided, then a longer distance covered would show motivation – aiming to improve on a target distance – had a stronger effect. My brain, my effort, would be stimulated purely by the desire to prove man could beat a machine.

I set up the machine for a third time. I managed 1,134 metres, which might look like a victory for the machine, but the unblinding of the test changed my tactics. With a visible target, I started off too quickly and ran out of energy after three minutes. That's my excuse anyway. And, of course, such one-off experiments prove nothing. To build a solid case I would need to repeat the routine dozens of times and then average out the results. I'll leave that to somebody else.

For endurance sports and the effect of cognitive

enhancement the early results are interesting, but there's a long way to go until scientists can be sure of a benefit. So, what about other mental skills in sport, those that rely on ability rather than determination?

William Stubbeman is a psychiatrist in Los Angeles. Tanned and fit, his easy demeanour hides the trauma he sees most days. Many patients view Stubbeman as their last chance. Mavis, for example, was sixty years old and had struggled with bi-polar disorder for most of them. She had been given the devastating shocks of electro-convulsive therapy a staggering fourteen times but with no benefit. If Stubbeman could not help her, Mavis had resolved to kill herself.

Colin had reached that stage already. Only nineteen, depression had such an impact on Colin's young life that by the time he walked into Stubbeman's clinic he had already tried to commit suicide.

Both Colin and Mavis walked away from his office, Stubbeman says, fully recovered, after he used magnetic brain stimulation to treat their conditions. That sounds extraordinary, but it's not the reason I arranged a Skype conversation with him. I wanted to ask about the impact his brain stimulation had on his tennis.

Stubbeman plays a lot of tennis and he has won a lot more matches recently. Much of the improvement is down to a staggering increase in the number of first serves he says he now delivers with unerring accuracy.

Impressed by the response to the brain stimulation he

was delivering to his most severe psychiatric patients, Stubbeman tried it on himself. He used the same kind of electrical brain stimulation as my kit to activate a brain region under his right temple – the right inferior frontal cortex – which is associated with the visual identification of objects. It's the same set-up as the US military used to help people find hidden threats. Stubbeman instead visualized a tennis ball, and hitting it to serve an ace, with the ball successfully landing in an imagined three-feet-square target inside the opponent's service box.

The stimulation was done before, during and after sessions during which Stubbeman hit dozens of first serves. The stimulation improved his serve accuracy by 20–30 per cent, he says. And the effect has lasted ever since.

He tried it on his tennis coach, a former professional. This time, the brain stimulation improved the serving accuracy by 13 per cent on the day; and by a whopping 22 per cent when they returned five days after the stimulation.

Stubbeman knows better than to publicly claim too much for the results of his experiment, presenting them only at a specialist conference, and arguing only the effect deserves wider study in larger controlled trials. As a scientist, he is cautious about the implications. But as a tennis player he says he is sure the brain stimulation has improved his game, and is responsible for him winning more.

Use of a performance-enhancing drug that claimed such a dramatic improvement would surely be banned. But at present, tennis players and anyone else who wants to are free to experiment with brain stimulation as much as they like. In fact, in 2016, a US company called Halo Neuroscience launched a high-end electrical brain stimulation device to encourage them to do so.

The company has packaged the battery and electrodes into a set of funky-looking headphones – no knitted Spiderman hat for them – and distributed them to elite sports stars and teams across the US.

The Halo kit targets the motor cortex and encourages athletes to use them as they practise a specific movement or routine. The company says this will help with motor learning, by making the brain neurons more likely to form the necessary connections. The US ski and snowboard team has been experimenting with the brain stimulators to train its ski jumpers to push off from the ramp. And it says it has seen visible and significant improvements in power output, as well as better control over technique.

Even the best can lose the firm grip they usually have on technique. The golfer Ernie Els has won sixty-odd tournaments, including the British Open and the US Open twice each – one of only six players to do so. He was the first to win 25 million Euros on the European tour and is a former world number one. He is heavily involved with autism charities (he has an affected son) and is generally considered an all-round nice guy. So it's unfortunate when you Google his name that among the top links suggested is a video clip of Els taking a swing at what is widely described as the worst putt of all time.

Some say it was six inches, others it was a full foot. Either way, it was a stinker. An unflattering camera angle from behind catches the full horror – the ball squirms almost sideways off his putter and doesn't even graze the edge of the hole.

This moment of ignominy for Els came in late 2015 at an event at Carnoustie, a notoriously tricky golf course on the blustery east coast of Scotland. In a later interview, an admirably upbeat Els tried to explain what went wrong, and offered a lengthy technical explanation of the weight distribution of his putter, how it hung in his hands and how he felt he was struggling to swing it hard enough to even hit the ball. All over a six-inch putt. He was, in other words, thinking about it too much.

To think too much about what you are doing is a cardinal sin in sport. From the footballer put clean through on goal with an age to determine what to do next, to the cricketer trying to remember to shuffle his feet, not move his head, swing his bat and keep his eye on the ball, as it bounces and skids towards him at near 90mph, a focus on thought rather than the instinctual appeal of action has been blamed for generations of high-profile failures on the sports field.

Some sports psychologists argue this collapse under pressure – choking – is inevitable because of the way sports technique is taught. Or because it is taught at all. Conventional so-called explicit learning – put your hands here, move your feet like that, keep your weight on your front foot – is vulnerable, they say, because it leads to conscious awareness of motor skills, and so produces conscious efforts to control what should be subconscious processes. Coaches call this paralysis by analysis.

The alternative is implicit learning, which lets people work out what to do, but without ever being able to explain it. The technique is worked out by their unconscious mind, which then makes it available on demand. Learning to ride a bike is the most common example of implicit learning. We have no conscious awareness of the physical tweaks and shifts in balance, for example, which keep us upright as we pedal along. Equally the best way to learn to ride a bike is not to listen to instruction but just to have a go: the unconscious mind gradually works it out and so you improve.

Implicit skills are harder to teach, because attention has to be deliberately drawn away from performance. Tennis players, for example, can be taught implicitly to read the direction of an opponent's serve by being asked to judge the speed and not the direction of the ball. In doing so, they learn to identify and act upon the visual cues that indicate direction, without knowing or being able to explain how they do so. Some coaches get basketball players to sing while they practise free throws, to take their conscious mind off the technical execution of the skill.

These implicit learning methods have one goal in common, to minimize the role of working memory, and so the scope for distracting recall. Electrical brain stimulation might offer a better way to do this. Rather than sideline working memory, why not try to turn it off?

It's too late for Ernie Els – the technical details of a putting stroke are seared into his memory. But if beginners could be taught to putt without explicit knowledge of club-head weight and swing speed, then would that help? Early results from a pioneering trial of brain stimulation on sports ability at the University of Hong Kong suggest it might.

Researchers from the university's Institute of Human Performance recruited twenty-seven students with zero experience of golf and gave them a crash course in learning to putt. Their improvement came from implicit learning: the students were left to figure out the best technique themselves, in a series of fifteen- to twenty-minute practice sessions. Each time, the students had to try to hole a six-foot putt. To make it easier, they hit the balls along straight and level patches of artificial grass with no slope or speed to judge.

While they learned the motor skills involved, half the students had their brains stimulated, but not in the usual way of making a targeted region work harder. This time, the sports scientists placed the cathode – the inhibiting electrode – over the left dorsolateral prefrontal cortex, an area above the left eye strongly associated with (among other things) working memory. The researchers wanted to use the current passed into the brain not to activate the working memory, but to turn it off.

The scientists brought the students back another day, and with no brain stimulation, asked them all to try the putts again. As the scientists expected, the students who had their working memory region inhibited by the electric current in the training sessions holed consistently more –

191

between three and five successful putts from seven – while those who did not receive the brain stimulation managed between two and three. (The students did not know if they received the stimulation or not.)

Their better putting performance, the scientists suggested, was down to a greater amount of implicit learning. Even though the researchers offered no explicit instruction in golf to the volunteers, they suggest working memory still interfered with learning and performing the task. So, switching it off, or at least turning down its power, helped the students learn.

The relationship between intelligence and the ability to learn is a complicated one. Not all learning, as we saw above, requires conscious thought and so applied cognitive power. And learning does not proceed smoothly. I experienced this in my treatment for OCD. Although I was learning to change the way I processed thoughts and handled anxiety, the results emerged in an unpredictable – what scientists call non-linear – way.

The dose was constant, three hours of cognitive behavioural therapy a week, but my response was haphazard, and the benefit – my reduced anxiety and freeing of thoughts – came in jagged peaks, leaps and bounds. I wasn't being treated, I was being taught. Like when I learned to ski, or tried to play the guitar. Hours of fruitless effort and then, oh wait, now I get it.

It felt like a phase transition, those tipping points of the physical world when small changes really do make a big difference. It's the kettle boiling the water, the steady input of heat lifting the temperature like a nervous opening batsman through the 80s and 90s until peeeeeeeeep, the magic 100 degrees is reached and all the extra heat in the world won't budge it higher. That water is going no hotter. All the effort now goes into changing to steam. The shift from 98 to 99 degrees takes as much dose as the shift from 99 to 100 degrees. But the response is totally different; from liquid to vapour and from anxious to calm.

A financial adviser once told me that almost all of the gains he made on money he had invested for clients over a decade came in a handful of days; those sudden spikes, the storms, when the effect is out of the control of the dose. Investors who constantly took their money in and out, he said, would miss out on those.

What if smart drugs or brain stimulation can help people's brains transit between phases, to find a way to shift cognitive performance to a higher level? That's certainly the hope of some psychiatrists using neuroenhancement as an adjunct for standard therapy. They want to see if chemicals or a tickle of current can help people to make the kind of mental transition needed to gain more control of harmful thoughts. For this kind of therapy cannot treat mental problems from the outside, it can only help patients find and unlock some cognitive skills already there.

Unlocked. That's how John Elder Robison describes the release of his emotional intelligence and it's how I felt when I started to make progress in my own therapy sessions for OCD. The new ability is not planted or encouraged. It is released; just as the steam is released from the water. And, experience shows us, there are many different skills and abilities – phase transitions – that can be released in the brain. We just need to find the right way to give them a little push. For experience shows that if we can dose the brain in the right way, the response can be extraordinary.

ELEVEN

The Little Girl Who Could Draw

When we returned from the holiday to France during which I followed Tom Simpson and cycled up Mont Ventoux, I asked my five-and-a-half-year-old daughter to draw a man on a horse. Ten minutes later, she proudly presented this:

Now, obviously, I consider my daughter to be an artistic genius. But I will agree all of her talent perhaps doesn't show itself here. In fact, as much as it pains me to admit it, the drawing is pretty average for a child of her age.

The scale is good – the rider's head, body and leg are in the correct proportions – and while the reins are a little wayward, the overall impression is definitely man-on-a-horse-y. But if you wanted to be critical you might point out the legs of the horse are rather stuck on to its body, and the leg of the rider is a bit of a token effort. But then, that's common with children's drawings. They show what they know to be true, not what they see. The same principle explains how children draw square tables. They know there is a leg at each of the four corners, so they draw all four legs in place, as if the table top was transparent glass.

On the next page, there is another drawing of a man on a horse, done by another five-and-a-half-year-old girl. It's a bit better. Actually, it's a LOT better. It's so good, in fact, most artists and child psychologists agree it should be impossible for a child of that age to draw. Indeed, the first people to see the picture insisted the girl could not have drawn it. But she did. Take a look.

The girl who drew this picture was called Nadia Chomyn. When Nadia's mother first showed it and others her daughter had sketched to clinical psychologists in Nottingham in the early 1970s, the scientists assumed the mother was mistaken, or worse, trying to deceive them.

The detail, the perspective and the unusual head-on approach were all the signs of a far more mature mind than

that of a five-year-old. Most strikingly – and this perhaps didn't register with you at first –the image breaks the boundaries of the paper. That is unheard of for a young child. Much older children and many adults strive to make a picture fit the frame, even compressing features and squashing letters as they approach the margin, so the image does not bleed off the edge.

Nadia was born in 1967. Growing up in the English Midlands she stood out, not least because her parents and grandmother were from Ukraine. Her father spoke good English but the rest of her family didn't. In fact, her grandmother rarely spoke at all, which might help explain why Nadia, who spent most of her time with her, was virtually mute. As Nadia grew into a toddler, her behaviour became difficult to control. She would run off to the park, where she seemed oblivious to traffic and other dangers, and, while quiet most of the time, Nadia was prone to outbursts and bouts of aggressiveness and screaming. Struggling to cope, her grandma increasingly confined Nadia to her bedroom.

At school her differences from the other children were even more apparent, she struggled to show even a passing interest in her surroundings and would stare into space and wander aimlessly around the classroom. A year later, her language skills had not improved and her increasingly anxious parents sought medical advice, including getting her seen at the famous Hospital for Sick Children on Great Ormond Street in London. These early medical reports noted Nadia's exceptional drawing skills, but only when she was assessed by child psychologists at Nottingham University was the true depth of her artistic talent revealed.

Still, it was not a promising start in Nottingham. Large for her age, Nadia was clumsy, slow and lethargic. While one of the psychologists showed her toys in a playroom, the other scientists spoke to her mother as they observed Nadia through a one-way screen. There was, frankly, no sign of the artistic ability the mother claimed for her daughter. It seemed impossible that the detailed and skilful sketches the anxious woman clutched and tried to show to the scientists with pride could have been produced by the chubby, brown-haired girl they watched through the glass.

Handed a thick yellow wax crayon, Nadia merely

scrubbed roughly at the paper to produce a tangle of scribbles. The mother, the scientists feared, had brought her child into what was supposed to be a safe and controlled environment, apparently concerned about her welfare, and then lied through her teeth.

Everything changed when they gave Nadia a ballpoint pen. The sullen little girl came to life, smiling and chatting to herself as she quickly and confidently drew her pictures. Cockerels, dogs, cats, a giraffe, pelicans, human figures and the occasional train filled hundreds of pieces of paper. Each was put together with precise and accurate movements totally out of keeping with the slow and ponderous way Nadia usually held herself and walked.

And then there were the horses – the glorious, dynamic, saddled and decorated horses and riders. Muscles bulged as one galloped; its legs in perfect unison and ready to reach for another stride. One bared its teeth from the page. They were, one of the psychologists would say later, like the sketches left by Leonardo da Vinci.

The scientists were dumbstruck. Everything they knew about children's mental and drawing abilities – and they knew a lot – said it was impossible. Nadia's use of perspective, shading and foreshortening were years ahead of children her age. She did not include token objects: no sun in the sky or trees in the background.

How could the scientists be so sure Nadia's drawing was exceptional? They had seen more pictures drawn by five-year-olds than one would ever want to. A few years earlier, they had been passed some 24,000 'pictures of mummy' collected by the *Observer* newspaper when it ran a children's painting competition. These researchers had looked through enough legs and arms stuck onto formless female bodies to think they knew what children were capable of.

Asked to draw a square or a diamond, young children tend to construct it from four separate lines, taking the pen away from the paper after each to reassess. Nadia completed a diamond with two movements. Her hand–eye coordination was extraordinary. Children usually draw with jerks and small movements, constantly checking on the progress of each stroke with their eyes and altering the direction of the resulting line as they go. Nadia drew with continuous confident movements, as if she trusted her hand to do what was required. (This was a girl who could not tie her shoelaces.)

If the line she drew was not right, she would draw another and another until it was. That's different to the way other children's drawings are influenced by the process itself – after they make a line, they use it as a cue to draw the connecting lines, rather than rely on a more accurate mental picture. All lines are anchored to the first.

Another difference was how Nadia had no interest in colour. Her pictures were black, white and grey. They were as stark and monochrome and cold and as reluctant to engage with easy appeal to outsiders as the little girl herself.

The psychologists saw Nadia for five months, but her behaviour stayed the same. Mostly, she seemed disinterested in the questions they asked and the help they wanted to offer. But, in time, things did improve. Aged seven, she started at a special needs school and became more sociable. By nine she was talking much more, and able to initiate requests and conversations, such as asking for a sticking plaster if she cut her finger. She seemed happier, but as other mental abilities began to unlock, so her drawing ability faded.

Her pictures started to look increasingly like those done by her friends and other older children. She started to notice the drawings of others and to copy them. She started to include childlike features such as token objects and her sketches lost the life-like quality that previously made them stand out.

It was clear Nadia would always struggle to survive without daily help. Even as a young adult she had no concept of money, and could not reliably feed herself. A place was found for her at a dedicated residential unit, where she would spend the rest of her life. In 2010, Lorna Selfe, one of the original Nottingham psychologists, tracked her down for a visit.

By then in her early forties, the Nadia who Selfe met with was again a virtual mute and struggled to use cutlery, preferring to eat with her fingers. She still had fits of rage and had smashed objects in her room, including televisions, so it was kept bare. Nadia, Selfe concluded, was now 'an unremarkable person with severe learning difficulties'.

While her childhood masterpieces were hung on the walls of the unit, Nadia took no interest in them and no longer picked up her previously precious black pen. If one was handed to her and a member of staff suggested she draw, Nadia would break it in two. And her talent had receded. Her most recent drawings, some done when she was in her early twenties, looked like the work of a five-year-old. One she sketched of a horse looked more like the one drawn by my daughter. After a short illness, Nadia died in October 2015.

Given Nadia's learning difficulties, her drawing skill was extraordinary but it wasn't unique. As an eleven-year-old at a special needs school, Stephen Wiltshire was introduced in a 1987 BBC television documentary as the best young artist in Britain. With learning difficulties and an estimated IQ of 60, Stephen became fascinated by London buildings he saw from the street and started to draw them with terrific precision from memory. He's now a professional artist with a unique approach: places from Istanbul to Singapore arrange a quick helicopter trip and then set Stephen up to draw their cityscape from his amazing memory.

How can a brain that struggles with basic functions like language and understanding social cues manage to dazzle in these ways? Where do these magnificent splinter skills shown by Nadia and Stephen come from? Neuroscientists aren't certain, but one popular explanation – similar to the explanation offered to John Elder Robison – is a mental rewire.

Nadia's learning difficulties suggest she was born with a part of her brain damaged or not functioning properly. And when that happens, the brain can set up a neural diversion and ask a different region to step in and perform the damaged area's tasks. Because different parts of the brain tend to specialize in different functions, they often try to perform the same task in different ways. And, sometimes, those different ways can bring about a massive improvement, a mental phase transition.

Some of the most striking differences in the way brains operate are between functions on the left and right sides. A bit like the myth we only use 10 per cent of our brains, the common belief in a clear distinction between left-brained and right-brained people is false, but projected from a foundation of truth. The left and right side of the brain communicate across the divide all the time, and much of what we think and do, especially the higher-level stuff, is partly controlled by both hemispheres. But there are certainly some skills and abilities neuroscientists have traced to one side more than the other.

The left side of the brain, for example, is more heavily involved in language, speech and some motor skills. The specialisms of the right brain are less verbal and based on spatial awareness, visualization and construction skills. The left brain has more to do with functions that are logical, abstract, sequential and symbolic, including stuff like speaking and reading. The right brain focuses more on parallel processing and intuitive problem-solving strategies.

This perceived contrast between the hemispheres of the brain is the basis for one of the most popular explanations for the artistic skills of people like Nadia and Stephen.

When the left side of the brain is damaged, the theory says, the right side either takes over some tasks and does them in a different way, or is given more freedom to express its own specialist abilities. In such circumstances, the right brain is freed from the inhibition and oversight of the left.

When the left side of the brain does not develop normally, a common result is autism. Nadia and Stephen were both diagnosed with autism. In fact, about one in ten people with autism show some kind of exceptional skill. These people are called savants.

Savants don't fit the positive manifold model of human intelligence very well. If they are supposed to have a general intelligence, well, nobody told them that. Their intelligence, or at least their mental ability, is anything but general. It is highly, highly specific. There are savants who can multiply 1,345,873 by 749,823 quicker in their heads than you can punch the numbers into a calculator. There are some who are unable to read but can recall extraordinary details of past events. Others could not spell piano but can play one like a virtuoso. Their abilities could reveal a secret of mental phase transition, and so of neuroenhancement.

Savants are rare, but Darold Treffert knows hundreds of them. Treffert is a psychiatrist in Wisconsin who has worked with and studied people with extraordinary mental abilities since the early 1960s.

Treffert has seen hundreds, maybe thousands, of patients over the years. But he still remembers some children he met on the first day on the job. He started work on a new children's unit at the Winnebago Mental Health Institute near Oshkosh. The unit cared for thirty kids, all of whom had severe disabilities. Many were mentally handicapped and had been diagnosed with autism. All had been hospitalized.

Most of the children were there because they struggled to look after themselves. They found even the most basic tasks – washing, dressing, and eating – difficult. Robert, for example, was mute and suffered from severe learning difficulties. Yet Robert could do an extraordinary thing. He could rapidly put together a 500-piece jigsaw, picture side down. He could scan the shapes, mentally construct the upside-down puzzle, and then assemble the pieces on the table, Treffert says, 'with the precision, motion and rhythm of a sewing machine'.

Arthur was a 'walking this-day-in-history almanac' who had a vast reservoir of facts about what had happened on any given date, and loved to quiz people about them. Knowing what was to come each morning, Treffert would try to read up the night before, but was still unable to answer many of Arthur's questions.

Henry had a different kind of talent. He could shoot baskets, from the free-throw line, with unerring accuracy. He had an 'obsessive-compulsive' routine, Treffert observed, putting his feet in exactly the same position and holding the ball in exactly the same way each time.

Finally, John knew every bus route in the city of Milwaukee, from start to finish, the entire public transport system. John liked buses. He would walk around the mental health unit with a cardboard model of the destination window on the front of a bus, and had a scroll of paper with all the stop and street names to display.

One summer evening in 1980, Treffert's daughter Joni came home bursting with excitement about a miracle. The miracle was Leslie Lemke, a mentally disabled boy, who had been born prematurely and suffered terrible complications. Before he was six months old both his eyes had been surgically removed.

Leslie's foster parents were told he would probably die, but they refused to give up on him, his mother especially. She taught him to swallow and strapped him to her own legs to show him how to walk. Aged eight or nine, she bought him a piano, and Leslie would place his hands over hers as she played.

One night, woken by music, Leslie's mother walked into the living room to find her severely disabled son playing Tchaikovsky's Piano Concerto No 1. The composition had been used as the theme music for a television movie – *Sincerely Yours* – the family had watched earlier that evening. That was, she said, the first time Leslie could have heard it.

Leslie became so good on the piano he was invited to play at a local high school, as part of Wisconsin Foster Parent Recognition Month. Joni Treffert was at the concert and, when she came home, she told her father how Leslie had played 'from memory all sorts of classical, religious and popular music like a skilled piano virtuoso'.

Also at the concert was a film crew from a Green Bay television station. Amazed at what they had seen, they wanted to run it past a mental health expert and so took the footage to Treffert, who explained Leslie was a savant. The story went viral, well, as viral as it could in 1980. By Christmas, much of the US was in love with Leslie and the great TV anchorman Walter Cronkite closed his CBS Evening News show that year with the words: 'This is a season that celebrates a miracle, and the story belongs to the season. It's a story of a young man, a piano and a miracle.'

Three years later, Leslie was one of three savants featured in an episode of the show *60 Minutes*. Watching him, 'with tears in his eyes', was the actor Dustin Hoffman. When Hoffman was approached about a part in the movie *Rain Man*, and was told the producers wanted him to play Charlie, the younger brother of a man called Raymond who was an autistic savant, Hoffman said that, no, he wanted to play Raymond.

Rain Man introduced savant skills to a wider audience. In fact, the movie had such an impact the ideas of autism and savant skills are now tied together in popular culture. But it's important to stress savantism and autism are not the same. Most people with autism don't have a special savant skill, and they and their families are sometimes distressed by the common assumption they should do. And not all savants have autism.

The unusual skills shown by savants, Treffert says, are islands of genius. That makes Treffert a James Cook-like explorer, discovering and mapping these islands, and recording the human life he finds there.

He now has a roll call of more than 300 savants from

around the world, both children and adults. Some of these people he has met and some he has only read reports on, in local newspapers and websites. He finds many of them when they write to him, after they pull up his name when they search the internet after hearing or reading about others with similar abilities.

The majority of the savants on Treffert's registry have autism – but some don't. And, while most of the savants have had their unusual skills since birth, intriguingly for the idea of cognitive enhancement, some haven't. These people are called acquired savants. Their mental talent emerged later in their life. Thanks to a mental phase transition, an extraordinary new ability inside their brain was unlocked somehow. And if it was lying dormant in those people before they realized it, then it could be inside you too.

In his short story, 'Funes the Memorious', the Argentinian writer Jorge Luis Borges described a man who became a savant when he fell from a horse and knocked his head. The man developed a memory so powerful, Borges wrote: 'The least important of his memories was more minute and more vivid than our perception of physical pleasures or physical torment.'

But this detailed recall came at a price. The man could not think in abstract ways, he could not join all the dots he saw together by a shared concept. Even the concept of 'dog' with all of its variety and dissimilar animals baffled him, because to him a dog he saw from the side became a different thing when it turned to face him.

'Funes,' Borges wrote, 'could continually perceive the quiet advances of corruption, of tooth decay, of weariness. He saw – he noticed – the progress of death, of humidity. He was the solitary, lucid spectator of a multiform, momentaneous and almost unbearably precise world.'

Since Borges wrote his story in 1942, science has discovered people who really do become savants after a bang on the head. In the summer of 2015 I went to meet one of them.

Pip Taylor believes her first miracle began on an Edinburgh bus. It was late summer 1994. The city was returning to normal after another festival season and Pip was on her way to work. The sun was already high over the glittering Firth of Forth, where the seagulls danced around the high poles of the suspension bridge that carried the river of cars and trucks toward the city from the ancient Kingdom of Fife. Pip did not know it yet but her life was in danger.

Pip was twenty-nine years old and working as a waitress in the staff canteen at one of the office buildings in town. She liked to let her mind wander on the short journey through the rush hour streets from the Morningside house she shared with two friends. On that morning she was sitting with her back to the window, so she almost failed to notice when the bus pulled up at her stop. She jumped to her feet, black skirt flapping, and, forgetting the seat was on a raised platform, Pip banged her head, hard, on one of the time-worn leather loop hand straps that dangled from the roof. Deep inside her brain, something stirred.

By mid-afternoon that day, the lunchtime chaos had passed and the cleaned plates had been placed back in their teetering piles. Pip had to finish one final task and was steering the hose of a vacuum cleaner through the legs of the empty chairs. Satisfied for another day, she straightened her back and stood. That's when it happened. A sagging bulge of capillary vessel inside her head gave way, somewhere above her left eye. A cerebral aneurysm had exploded inside her brain. It produced a subarachnoid haemorrhage.

Pip did not know this at the time. All she knew was pain, and an odd sensation some water had been spilt onto the back of her head and was dripping down her neck. It was not water, but blood. And the blood was running down the inside, not the outside, of her head. A colleague asked Pip why she was crying. Until then, Pip had not realized she was.

According to the NHS website: 'A subarachnoid haemorrhage causes sudden, severe head pain. This condition requires immediate medical care to prevent brain injury and death.'

Her colleague asked Pip if he should call a taxi to take her home. Pip's reply almost certainly saved her life. She asked for an ambulance. Her survival, one of the doctors told her, was miraculous.

Today, Pip lives on the Wirral peninsula in northwest England. Head north from her house and you can catch a ferry across the Mersey. Go south and you wade into the silt of the River Dee. Nobody ever wrote a song about the ferries that cross the Dee, but the sluggish Dee tide rolling along the southern edge of the Wirral draws some tourists of its own. They watch as barges carry the giant wings of the Airbus-380 along the Dee to a deep-water port at Mostyn from the factories making them at Broughton. The town lies just to the west of Chester, which some historians claim the Romans had once earmarked for their capital ahead of London. Within the city walls of Chester, Pip Taylor was granted miracle number two.

It happened late one spring evening in 2012. Pip and a group of friends were at one of the city's pubs. Pip was talking to a man she knew. She was pleased to have bumped into him again. He kissed her and got up to buy some more drinks.

There are two versions of what happened next. In his, he returned with his pint of lager, her glass of white wine and a bag of cheese-and-onion crisps to find the woman he thought he had been getting on with had vanished. Still, if he felt his night was ruined, then it was nothing compared with how the evening ended for Pip.

Exactly what happened is unclear. Pip certainly can't remember, but friends, her sister and a doorman who was working in a rival pub across the street have helped her to piece it together. Smiling as he walked away, she stood, perhaps to stretch her legs, and took a step forwards. Then she passed out and collapsed head first down the set of narrow and steep stone stairs to the road. As she descended, she banged the right side of her head, hard, on each step. If she was still conscious at the top of the stairs, she wasn't at the bottom.

Another ambulance ride followed, and another stay in hospital; another trip home with a headache and instructions to take it easy. Against the odds, another doctor told her there seemed no lasting damage.

If anything, Pip felt better after the fall. Though she had recovered from her haemorrhage almost two decades before, it had left her demotivated and listless. It also prompted dizzy spells and the occasional blackout. That might have been why she fell. This time the repeated blows to her head, she says, seemed to put right whatever the haemorrhage had damaged and her mood in the weeks after the fall improved.

Something else changed too. As the time since her accident passed, she developed a craving. Her recovering brain fizzed with tension. She felt the urge to express herself. This wasn't out of character, Pip had enjoyed art at school and could draw a decent cartoon Snoopy. But her talent had been limited, so much so an art teacher gently advised her to look elsewhere. It was the eyes, always the eyes. She struggled to draw them. When she tried to draw real faces, the eyes would still always look like they came from a cartoon.

Remembering the art teacher's judgement, in the weeks after her accident Pip turned instead to wood carving. No use. She bought a heavy-duty bench and vice and installed it in the shed in her small garden. Three years later, it still stands there unused. Woodwork was too slow, she realized, and would not hold her attention. Modelling clay was the same. So one afternoon she picked up a pencil and notepad and started to sketch.

The result, a still-cartoonish fat boy, was good – better than she remembered. Startled, she drew a cat, copied from a picture in a book. That too was good. Something had changed. Pip's brain had changed. She had talent, skill, a whole new ability. She showed her mum. It was, her mum said, a miracle.

Judge for yourself. Before the fall, this is how Pip drew a girl's face.

And this is one she drew afterwards.

Psychologists describe Pip Taylor as an acquired savant. And reports of the mental phase transition in acquired savants like Pip Taylor often prompt two different reactions. There is wonderment, awe and mystery. And scepticism and disbelief. In many ways, scepticism and disbelief is the more rational response.

It's possible, of course, Pip is simply having us on. Perhaps she was always able to draw and her claims about the miraculous transition are a hoax. The rest of us, after all, have no way to check she was as rubbish before as she says. That was part of the reason I wanted to meet her.

Having done so, I believe her story. I can't offer a cast-iron guarantee the friendly and open woman in her early fifties who I shared a coffee with – whose mum lives next door and who sticks her head over the wooden fence between their gardens to see who she is talking with so early on a Monday morning – is not an elaborate fraud. But it seems unlikely. Not least because it's difficult to see what she would get from such a scam. She does not sell her pictures; she does not even like to give the originals away. And she is not a publicity-seeker. I only found her because a chance conversation she had with a head-injury charity about her drawings made its way through the media food chain from a local paper to the *Daily Mail*. When I arrived at her house for the arranged interview, Pip had forgotten and was still in bed.

The bang to Pip's head – and the subsequent phase transition in her brain – improved her drawing ability. But did it increase her intelligence? She reported no other obvious changes in mental ability, such as maths or memory. And without the results of an IQ test before and after the incident we can't judge that. But according to some measures, she did get smarter. For one of the skills long used to judge intelligence is drawing.

For decades, psychologists have tested children's mental development simply by asking them to draw a person. Points are awarded for life-like features, such as the number of arms and legs (and other body parts) depicted, and if they are attached and presented in the correct proportions. A nose, for example, must be drawn longer than it is wide. Clothing scores points, as long as the body parts do not show through. Extra points are awarded for details: eye-lashes, pupils, toes, thumbs, beards, teeth and recognizable hair styles.

Similar to Alfred Binet's tests of mental age, the draw-a-person exercise assesses stages of development in kids' artistic ability. Aged two to five years, children usually draw faces only and then people as tadpoles: heads on legs. Then come bodies, with arms attached halfway up, and then hands with a generous supply of fingers. From four years, the drawings start to feature details such as a waist, and while the arms correctly move up to the shoulder region, legs are typically set well apart and parallel. From five years old, pattern and decoration feature, and the neck arrives. A year later, the figure is often placed against a (often colour-ful) background, and aged eight, the child starts to draw dynamic scenes, like people engaged in an activity.

Drawing gets more difficult at this point, and decoration and colour are not enough to make a picture look good. Human figures in action demand that arms and legs bend and interact with objects in a realistic way. Combined with increasing self-consciousness, this is a key stage in development. Unsatisfied with the results, some children give up drawing. Others are able to upgrade their instinctive skills to conscious technical ability. Many (and this was me) retreat into facetiousness, and realizing they lack the ability to progress, deliberately draw what they know – and can defend from criticism – as unrealistic images. Bodies are distorted, often to bizarre effect. Jokes and visual puns appear. The child wants their pictures to be admired, not as artistic, but as clever.

Intelligence tests based on drawing a person simply compare a score for a sketch to the expected score for a child of that age. According to this system, the young Nadia Chomyn had an intelligence off the charts.

Scores significantly lower than the average on this drawing test signal a child of equally low intelligence, complete with all the social and educational implications that such a designation has brought over the years. The test is rarer now than a few decades ago, but it's still used, particularly in some developing countries. It's even used in some places to probe the intelligence of adults, typically those with learning disabilities and other mental disorders. And that means that bangs on the head, or other techniques that can improve drawing, can be argued to increase intelligence. And one way scientists have found to improve someone's drawing is electrical brain stimulation.

In 2013, scientists at Harvard University used electrical stimulation to improve the drawing skills of a former builder called Bob. Bob had suffered a stroke on the left-hand side, and the scientists wanted to see if they could encourage the kind of rerouting of activity in the brain believed to unlock new abilities following such trauma. So they attached an electrical brain stimulator to the skull above his prefrontal cortex on the right-hand side. They asked Bob to draw pictures: sometimes while they used an electrode to stimulate his brain, and sometimes when they only pretended to.

We're back to horses. Here's Bob's 'before' picture, without the electrical stimulation:

And here's the drawing he did when the current was switched on:

Admittedly, neither picture is striking, especially when compared to Nadia's and Pip's. But then Bob only had two and a half minutes to draw each one. The scientists didn't want to expose him to the electrical current for too long. Nadia and Pip could, and did, take hours over their pictures. Anyway, it's not their general artistic merit that matters, but any notable difference in quality between the before and after sketches.

Bob also drew houses. Here's one without the current to his brain:

And here with the current on:

The Harvard scientists mixed the pictures up and asked eleven of their colleagues to rate them from one to ten on the following categories: creativity (use of the imagination or original ideas), perspective (representation of solid object on a two-dimensional surface), aesthetics (beauty or the appreciation of beauty), reality (representing a real thing, not imagined) and accuracy (careful and free from errors).

In each case, the scores for the pictures Bob drew under the influence of the electric current were significantly higher. Again, I think it's clear that Bob did not become an artistic genius, but, according to the results of the study, the electrical stimulation had a significant effect.

To understand what is happening during this type of change in the brain, how it shows itself in savants like Pip Taylor, and how it offers a possible route to mental phase transitions and cognitive enhancement for all, we need to consider one further definition of intelligence. In 2014, while grappling with the problem of how to develop artificial intelligence, computer scientists from Israel and the US posed a provocative question: 'How much information should we drop to become intelligent?'

While intelligence is usually measured in additive terms – the more factual knowledge the better – the scientists argued true intelligence demands some of this knowledge be ignored in favour of higher-level abstraction. Cognition, they say, is categorization, and has to involve the loss of some of the concrete detail. What's the difference between concrete and abstract categories systems? The mental process of abstraction and categorization sees two, 2, II, ii and deux as all the same. But in concrete terms the order and shape of the marks on the page are different.

As children's brains develop they move from concrete thinking to abstract. They start to ignore, for example, the differences between individual cats and instead group them into an abstract category of Cats, which are different to Dogs. We have to do this or our brains would be swamped: imagine trying to hold the detail of every type of cat, dog and an infinite number of other concrete examples in your consciousness. Much easier to think of a Cat by default, and then zoom into a specific type of cat if necessary. For this kind of dropped concrete detail is not lost: it's buried in our memory banks should we need it, and more importantly, should we go looking for it.

To achieve their mental phase transition, many savants seem able to access this dropped concrete information. But in exchange, they lose the ability to process it into abstract ideas and concepts, to make the most of the P-FIT mental circuitry that helps humans interact with the world. A similar trade-off seems to take place in autism. One of the most popular explanations for the way the autistic brain operates is called the weak central coherence theory.

A rewiring of their brains, the weak central coherence theory says, makes people with autism unable to perform the higher-level processing to convert concrete information – the where and what – into abstract concepts, the why. Someone with this kind of mind could easily tell you how they keep fit – running, swimming, whatever – but might struggle to explain *why* they do so.

Again, drawing and art helps to show this distinction between abstract and concrete most obviously. One group of acquired savants are people with frontotemporal dementia, a relatively rare debilitating condition linked to Alzheimer's disease. It usually emerges earlier in life than other dementias and it attacks the frontal and temporal regions of the brain we rely on for language, to plan and make decisions, and to govern our behaviour. As these neurons, and the connections between them, are slowly poisoned by invasive, toxic, sticky clumps of protein, they die and their functions wither.

Patients with frontotemporal dementia often lose the ability to speak and comprehend, and combined with a decline in other social skills, they can revert to a childlike state. But they can also show one or two mental skills that continue to shine amid the ruin caused by the disease.

Mavis, for example, spoke eight languages and played professional-level bridge. She noticed the first symptoms of dementia at sixty-four, and when she had her IQ tested aged sixty-eight it was obvious some parts of the brain were being more affected than others. Her arithmetic was brilliant, but she often couldn't remember a single word of a list read to her. She lost her general knowledge and struggled to answer basic questions, but could still play chess.

In some extraordinary dementia cases, as the sufferer's brain dies off a different skill can emerge. Patients with frontotemporal dementia can feel a compulsion to draw and paint pictures, and many of them do so with a savant-like skill or style they did not show before their illness.

The neurologist Bruce Miller has studied this transition in a group of dementia patients at a hospital in San Francisco. He believes – and brain scans support the idea – that damage to the front left part of the brain triggers a burst of activity on the right. And with the left side out of action and the right side in charge, the pictures tend to be focused and realistic, without significant abstract or symbolic features.

The dementia patients, in other words, showed a switch from abstract categorization to distinct concrete detail. The pictures they produced were technically proficient, but limited in approach. They were not what artists describe as conceptual or abstract works. Indeed, a few cases of established artists who have suffered this dementia show a similar trend, away from abstract shapes and more towards conventional images of landscapes and portraits of people and animals.

Research with savants is rare, but some studies show evidence of rewiring in the brain that would support such a shift in emphasis. In 2014, Japanese scientists reported research they carried out on a retired office worker called JN, who suffered a brain haemorrhage in his mid-sixties. JN liked to paint, and one of the last pictures he completed before his brain injury was a portrait of his wife. His brain damage was to the left prefrontal lobe, the same site as injuries in the frontotemporal dementia patients, and the scientists were keen to see what would happen to his art.

About a year later, with no prompting from anyone, JN reached for his paint brushes again. One of the subjects he returned to was his wife, and he painted what looked, to the inexpert eyes of the researchers, to be a more realistic and life-like portrait.

To check, they brought in the acknowledged specialists – twenty-seven professional reviewers from the Tokyo National University of Fine Arts and Music. Without telling them why, or the circumstances, the scientists asked the critics to judge and score each picture on several criteria.

The results confirmed first impressions. The experts rated the second picture as higher on realism and technical skill, but lower on aesthetics and evocative impact – exactly what would be expected as JN's brain switched from abstract to concrete processing.

Scans of blood flow in his brain showed the back of his parietal lobe on the right-hand side had become more active. The injury to JN's left-hand prefrontal lobe, the scientists said, provoked a compensatory increase in the activity of this other region. The inhibition of the left was lifted and the right side of JN's brain was thrown the keys.

Beyond art, another good example of this distinction between abstract and concrete information is the savant skill of absolute or 'perfect' pitch, the ability to recognize a single musical note played alone. Only one person in every 10,000 has absolute pitch and it is said to be impossible to teach. Most musicians – even those we think of as geniuses – don't have it. They need a reference note to work from. Play them a C (and tell them it's a C), then they can correctly call everything that follows – G, E, F, whatever. But play them a C without identifying it for them, and someone without absolute pitch will struggle.

Yet arguably all of us have the mental equipment for absolute pitch. To hear sounds we must analyse and identify the discrete frequencies of all the component noises. This means, in a physiological sense, our brains should be able to identify a sound with a frequency of 440 hertz as the note 'A' just as we identify light with a frequency of 660 terahertz as 'blue'. Yet most of us can't or don't.

This seems to be because our brains usually pass over the information on the individual sounds and instead focus on the combined effects of all the different notes, and the relationships between them, because that is the aspect of what we hear that we consider most important. In contrast, savants with perfect pitch seem to have access to the raw, unprocessed data – the concrete information on the sound frequency.

(Something similar seems to happen when some people with autism look at fluorescent lights. While other people see the big picture, the continuous glow, the unusual, concrete-focused processing of the autistic brain can see it flashing on and off 120 times a second, which can explain why some people with autism find them so disorientating.)

Another group of people with unusual mental abilities supports the idea that a switch from abstract to concrete processing in the brain can trigger high-level savant cognitive skills, and perhaps offers a route to neuroenhancement. Just like savants, these people seem to have privileged access to cognitive machinery denied to the rest of us. They have synaesthesia, an unusual condition that blends and confuses the input and output of different senses. Some people with synaesthesia see sounds and hear colour; those are probably the most well-known forms of the condition and have received wide attention.

Lesser known is a form of synaesthesia where people see the passing of time in physical space. We all do this to an extent. In cultures where reading and writing goes from left to right, people tend to think time flows in that direction as well. Ask someone what they did last week and they are more likely to wave around their left arm. Future events are denoted with the right. In scientific tests, people respond to the names of the earlier days of the week and earlier months of the year when allowed to signal with their left hand, and to the later days and months with their right. American children order time-related concepts like meals from left to right (breakfast on the left and dinner to the right) while Arabic-speaking children place them right to left.

In a similar way, we tend to visualize smaller numbers on the left. We hold an imaginary line in our mind's eye onto which we hang time, numbers and other sequences. Psychologists call this process the reification of abstract concepts into concrete representations. It reverses the process of how intelligence demands concrete examples are categorized into abstract concepts. It turns the abstract concept of time in the brain into mental units as discrete and concrete as the numbers on a clock face. This type of concrete processing notices the counting of minutes and hours and days, rather than the sense of passing time. And this sense of detail in time is what some people with synaesthesia report.

Most striking, they say they are consciously aware of their imaginary time line, which they clearly visualize in three-dimensional space, either in their mind's eye, or circling or wrapping around their body. Just as an architect might design a building in their head – this window over here and this bit of the roof at a different angle – so these people construct a spatial framework for time – early morning over there and late afternoon up a bit. It might sound unusual, but to a visuo-spatial synaesthete, as they are known, to *not* visualize time in this way is equally bizarre. That's a common feature of many forms of synaesthesia; people with the condition are often astonished to discover (and some discover it late in life) not everybody experiences the world the way they do.

The concrete visualization of time was noted as long ago as 1880 by the early intelligence researcher Francis Galton (he of the ugly map and eugenic ideas), who made woodcuts and engravings to reproduce the mental maps his subjects reported. One, a 'mathematical astronomer of rapidly rising reputation', told Galton: 'The numbers 1, 2, 3, 4 etc are in a straight row, and I am standing a little on one side. They go away in the distance so that 100 is the farthest number I can see distinctly. It is dusky grey and paler near to me; up to 20 it occupies a disproportionate size. There are sorts of woolly lumps at the tens.'

Recent research at the University of Edinburgh suggests visuo-spatial synaesthesia is more than a mental curiosity. It could confer cognitive benefits, some of which seem to relate to the kinds of questions asked on IQ tests. It could, in other words, increase intelligence.

In one Edinburgh study, ten synaesthetes consistently scored better than other people on tests of their awareness of time. They were significantly better, for example, at identifying the years in which major events of the twentieth century occurred, when Oscar-winning movies were released, and when songs reached the Christmas number one slot. That might not seem surprising, given these people's brains focus so heavily on time and its passing. But the study showed the benefits went further.

The scientists also gave the synaesthetes tests to measure their spatial ability, by asking them to manipulate real or imagined objects in 3D space. They were shown images of complex shapes and told to build them one-handed from wooden blocks. They were asked to identify objects (like a gun and a trumpet) placed at various angles from their silhouettes. And they had to work out how a mixture of blocks would look when rotated. They performed better in all of them.

The above-average mental skills of the people with synaesthesia were especially noticeable when the scientists asked them to recall autobiographical details. Each was given a series of years, spaced evenly from when they were five years old until three years before the test, and told to write as many facts as they could remember. They were given one minute for each year – what they were doing, who their friends were etc. Most were aged in their thirties and forties, so the details they were being asked to remember were from two or three decades before.

All of the synaesthetes scored impressively, but the recall of one, a thirty-two-year-old man called Ian, blew the scientists away. The synaesthetes, on average, could remember seventy-four facts from the nine selected years combined, almost double the number detailed by other 'normal' volunteers (try it, it's harder than it sounds). But Ian wrote down 123.

Just like the way the Nottingham psychologists re-sponded with disbelief to the claims of Nadia's mother that her daughter had drawn those astonishing pictures, the Edinburgh scientists struggled to believe Ian's score. So they checked if he was doing what he remembered during those years with his sister, his aunt and his fiancée. Everything happened as he remembered.

The cases we have discussed – autistic savants, dementia patients and synaesthetes – together demonstrate how the human brain can break from the collectivism of general intelligence. The performance of one mental skill can vastly overshadow the rest. And this imbalance seems to be down to changes in the way the brain communicates, the neural routes selected and the regions brought online. This altered brain state arises as compensation for loss to disease or as a consequence of unusual development. And importantly, in the search to harness this effect for cognitive enhance-ment, experiments show these positive effects can be triggered and released without the downside of damage and disease.

TWELVE

The Genius Within

For the last century or so, doctors thought the brains of savants were alien, and their feats of marvel were off-limits to the rest of us. Savants were simply born that way. But the transition of the dementia patients and others suggests that might not be true. Their change came on as parts of their brain were turned down. They did not learn to paint and draw in such strikingly different ways, their ability to do so was released from inside. Brain scans of acquired savants seem to confirm no idle region of their brain springs into life, no part of that apocryphal unused 90 per cent of the brain holds their secret. We all have the same equipment; it's just some people use it differently. And, handily for neuroenhancement, this hidden brain equipment can do much more than paint and draw. Many other savant skills can be released by a bang on the head.

Orlando Serrell, for example, was ten years old and playing baseball with friends when, racing towards first base, he felt a flash of pain and fell to the ground. Flung by a playmate, the solid ball had struck him high on the left side (the left, again) of his head. Life changed for Orlando that day. It became a lot more memorable.

Orlando had a bad headache for days and, as it eventually subsided, he developed a powerful memory and could recall with remarkable detail the events and weather of every single day since his accident. And something else was different. The youngster found he could identify the day of the week from any date.

The effect continues today. Throw a random date at him, as long as it was after when the injury occurred – 17 August 1979 (a Friday) – and Orlando can reply instantly with the day, and with a moment's thought, the weather. It's a skill called calendar calculation and one Orlando says is neither welcome nor any effort to perform. He does not know how he recalls the information, he says, it just comes to him. And he insists he has not studied and learned the details he can recall. He has better things to do, he says, than spend hours looking at old calendars and weather reports.

Then there is Louise, an American woman who fell heavily while skiing on a slope 'covered in moguls in fading flat twilight' and broke her collarbone and banged her head. She was diagnosed with moderate concussion and, she said, 'over the following weeks, shit got weird'. Louise found she could remember too much. She could recall and recreate with extraordinary detail the floor-plan of every building she had ever been through – the rooms, doorways, corridors, everything.

'This applies only, however, to static spaces. I can tell you what was in the vending machine at the rest stop off the

530 between Little Rock and Texarkana, because it's a fixed point. I remember what was in there when I was there. It's because of this that I still (much to my chagrin) will put my keys down in the wrong place, and forget where they are, since they are no longer at their fixed point.'

These examples fit the pattern: some kind of damage rewires the brain and triggers a more concrete-based processing. But they also introduce something new: the ability to recall concrete details stored in the brain from years before, which would never otherwise have come to conscious attention. They were buried in the subconscious, and it's the subconscious that seems to offer the secret to mental phase transitions.

The only way we can explain Louise's knowledge of all those buildings and Orlando's accurate recall of weather from days long gone is that their brains had stored the information somewhere without their knowledge. And then the blow to their heads opened the store and made its contents available to their conscious mind.

We all have a subconscious store of accumulated concrete information – names, faces, and the answer to a crossword clue on the tip of your tongue released only when you think about something else. The subconscious can sound mysterious, and in some ways it is, but there's nothing unusual about the brain storing information, even important information, in there – it leaves conscious attention free to attend to more urgent matters.

Some of this concrete information buried in our subconscious we have deliberately learned, from multiplication tables to the dates of battles and birthdays. But some goes directly from our senses into our subconscious without us ever being aware of it.

Our senses, after all, are constantly bombarded with sights, sounds and sensations and, to stay alive and thrive, it's important for the brains of all animals to be able to filter these stimuli, to discard the irrelevant and bring to attention the most important. That can't be done instantly, so the brain needs a way to store information while it sorts through it. That's called sensory memory, and it tends to hold stuff for just a few seconds. The most obvious example is when children swirl lit sparklers around in circles at firework displays. The trace of light you see left behind isn't there – it's held and presented for your attention by your sensory memory, but only briefly. Most of the information put into the sensory memory is discarded without us ever being consciously aware of it. But the experiences of Louise and Orlando suggest that some of it gets filed away somewhere, and that it can be retrieved.

When I tell people I am writing a book about how a bang on the head can release mental skills from the subconscious, and how scientists are seeking ways to improve intelligence and cognitive abilities, almost all of them mention to me the same example: they heard about or watched a programme on a man or woman who wakes from an accident and can speak a foreign language.

Sadly, that kind of change is impossible. As mysterious as sudden gains of ability in the brain can appear, they must remain rooted to reality and personal experience. Detailed knowledge and expertise of a previously unknown subject, like a foreign language, can't simply spontaneously appear, or be planted, in someone's brain. If I have never studied it then a bang on the head won't make me be able to speak Russian, any more than it could grant me the skills and knowledge to fly a helicopter.

But it could make me *sound* like I speak Russian, and this is almost certainly where the stories come from. It is not a foreign language people acquire in these circumstances, but a foreign accent. This strange switch was first identified more than a century ago when the French neurologist Pierre Marie encountered a stroke patient from Paris who started to talk like he came from the Alsace region. Dozens of similar cases have been reported since, from Americans who talk 'posh' like the British, to a woman from Portsmouth who picked up what sounded to her friends like a Chinese accent. It's called Foreign Accent Syndrome and despite the air of mystery with which examples are presented in media stories, there is usually a mundane explanation.

The answer is not why the new accent emerges, but how it does. Neuroscientists and linguists think the accent is a speech impediment. When most impediments emerge, often after a stroke, the speech is often slurred or broken. In some cases the changes in emphasis – cadence, pitch and rhythm – is mild and specific enough to sound to us like those of an existing accent. But this is a total coincidence

– the connection to the foreign country is made in our brains, not in the speaker's.

In a related syndrome, a recovering person does start to speak a different language, but the explanation here is equally plain. In these cases, the speaker knew the language already, but spoke it less often or had not done so for some time – rarely enough at least for those around them to be quoted as 'baffled' about the new, and to them incomprehensible, vocabulary the affected person comes out with.

New abilities to speak foreign languages may be a red herring, but there are plenty of just as bizarre, real, and well-documented examples of novel skills, sensations and behaviours that do emerge from the subconscious brain, usually when it's under some kind of stress. Some of these offer some clues on how extra intelligence might be found and accessed. They're called experiential experiences. And most of us have some experience of this type of glitch in the matrix. Probably the most widely known experiential experience is the feeling of deja-vu.

Science on experiential experiences sometimes gets hijacked to promote the claimed existence of spiritual and religious experiences. But the literal meaning of the word experiential – learned through observation and experience – shows what scientists think is going on. From time to time, stuff stored away in the brain's archives and data banks, like ancient memories and sensations, forces its way to the front of the mind. Given those archives can conceivably contain anything we have seen, done or thought, experiential experiences pulled from the contents of even the most average of lives can get pretty weird. That's the case with one of the most well-recorded examples of an experiential encounter: the near-death experience.

The near-death experience is a good example of the way the study of a genuine psychological occurrence has been commandeered by those with a religious agenda. People recover from near-fatal accidents and surgery with stories of seeing angels, for example, and these stories are pounced upon and celebrated by groups who want to believe – and want others to believe – angels are real and waiting for them.

Because of this, reports of other, what may well be genuine, near-death experiences tend to be dismissed, or at the least, viewed with hyper-scepticism. Talk of tunnels of light, voices of deceased relatives and feelings of calm, which would not raise a single sceptic's eyebrow if reported in a dream, are bundled with claimed glimpses of heaven as fiction when someone talks of them after they pull back from the brink of dying.

Yet, stripped of the religious interpretation, many reported near-death experiences make rational sense, especially given the chaos unfolding in the brain at the time. Indeed, one is so commonly reported it has become a cliché: the 'my life flashed before my eyes' moment.

The nineteenth-century Swiss geologist Albert Haim once had a severe fall while climbing a mountain. Convinced at the time he was going to die, he said later: 'I saw my whole past life take place in many images, as though on a stage at some distance from me. I saw myself as the chief character in the performance.'

Scientists call this a life review. When people report a near-death experience, they typically say it included a life review, which can appear to them as a movie, a series of still images or even just one or two flashes. The events can flow in chronological order, or run in reverse and finish in childhood. The life review in a near-death experience is a quirk of memory, an unprompted retrieval to consciousness of specific incidents and events, often for the first time since they occurred for real decades before.

Intriguingly, a similar sensation to the life review is described as a symptom of their condition by many people with epilepsy, often during the 'aura' phase of increasing electrical activity in the brain before a seizure takes hold. These flashes of memory can be terrifying and distressing, partly because they seem to emerge from nowhere. One epilepsy sufferer describes them coming after a seizure:

> I have strange memory flashes for a few days. Random memories, sometimes from long ago, relating to nothing that I am doing, saying or even thinking will just flash in my mind out of nowhere, like a three-second video clip.

Another says distinct memories signal a seizure is coming:

> If I hear the theme to *The Brady Bunch* or *The Jetsons* all garbled, with the smell of pizza from the lunch room in

second grade, I'm going down. Light distortion and Peter Brady. I hear teachers. Why? The pizza taste overwhelms my mouth and I stiffen like a board. Only one side of me, my stomach flips like I'm on a rollercoaster and then I go limp while exhaling. Then black out. If it's just an image flash, somewhere I've been as a kid, usually during second grade, I'll get sleepy standing up . . . My emotions can change with the memory. Deep sadness, my granny's funeral, like I'm there. Happiness too.

These retrievals of memories and images in patients with epilepsy do not have to come at random. Scientists have found a way to prompt them with electrical stimulation.

As part of surgical treatment for epilepsy, neurosurgeons implant a series of thin wires deep into the brain, which can be used to pass small electric currents into the surrounding tissue, with the resulting effects mapped using a series of conducting pads on the outside of the skull.

A man in his thirties was having this done by neuroscientists in Philadelphia when, as they pressed the switch to stimulate his brain, he told them: 'I'm remembering stuff from high school . . . Why is this stuff suddenly popping into my head?' It happened every time they sent the current into that section of his brain, including when he came back to see them two weeks later.

Brain surgeons at the Johns Hopkins Hospital in Baltimore had a similar experience with one of their epilepsy patients. In his case, the scientists could bring to his conscious mind one of four separate memories, depending on the location of the electrodes in the man's head they turned on. One pair of electrodes triggered the theme song from *The Flintstones* television show, which he had watched as a child. A different electrode pair brought back the memory of baseball commentator Richie Ashburn (who died in 1997) reporting a Philadelphia Phillies game, while another made him hear a familiar female singer, who he couldn't name. The final electrode pair triggered a strong memory of the Pink Floyd song 'Wish You Were Here'.

These are not one-off experiences. A large study by scientists in the south of France showed that of 180 patients with epilepsy given brain stimulation there in the early and mid-1990s, sixteen of them reported a side effect the researchers defined as 'dreamy states' – including the vivid recall of memories. When these people did have a flashback, it was not always welcome. One woman had repeated memories of a gas mask used to knock her out when she had her tonsils out aged fourteen. She saw a bald man dressed in black coming towards her and felt she was going to die.

In the last decade or so, psychiatrists and neuroscientists have started to use deep electrical stimulation to treat more conditions (in areas beyond the reach of DIY brain stimulation, such as the nucleus accumbens). It's used mostly to control the symptoms of Parkinson's disease. But, given the lack of options to treat many psychiatric disorders, it's increasingly used to tackle conditions like depression and obsessive-compulsive disorder. And this electrical stimulation has also produced some odd side effects.

240

Clinical reports of such side effects show a huge range of human observations and experiences can be dragged from the brain with deep brain stimulation: memories and the sensation of reliving a past experience, a feeling someone is nearby, mirth and laughter, uncontrollable crying, tastes, smells, warmth, chewing, bliss and increased motivation. All can be turned on and off with an external switch.

Just like the novel skills shown by the acquired savants like Pip Taylor, these sensations do not seem to be planted or introduced by the current, but released and activated. And besides memory and recall, other changes triggered by deep brain electrodes closely mirror those seen naturally in patients who suffer brain trauma. They can produce speech changes similar to those in foreign accent syndrome. A patient with obsessive-compulsive disorder in Holland given electric current to the brain shocked his wife when his voice took on a 'very distinguished' pronunciation and he began to use 'unusually distinguished' language. He started to ask for the public toilet rather than the loo.

A second OCD patient in Holland went in the opposite direction. His deep brain stimulation made him adopt, for the first time in his life, the accent common to his local region, and when the current was flowing he would, to his and everyone else's surprise, swear and use coarse language.

Perhaps most dramatically, neuroscientists have used deliberate electrical brain stimulation to make people involuntarily toggle between different languages as they speak. One man started to count in French and then – zap

– he continued in Chinese. Another Frenchman could be made to switch to English and then – zap – back again. Neurosurgeons in Italy made a Serbian woman talk in Italian to them. (All of these patients, of course, already knew how to speak the second language. Again, the behaviour is not introduced by the brain stimulation but released from the subconscious.)

The subconscious is more than a passive store of information. It has its own processing power that can be harnessed. Psychologists have long argued about exactly what this subconscious processing is capable of. They disagree over the relative ability of the conscious and unconscious mind to make decisions, for example.

One thing they thought they could agree on was the subconscious was less advanced than conscious processing. Non-conscious thinking could respond to stimuli, recognize objects, carry out familiar movements and recall basic facts. But more complex mental processes – planning, logical reason and combining ideas, the hallmarks of intelligence – were believed to require attention and so conscious thought. The conscious mind was smart and the subconscious mind was dumb.

That view was challenged in 2012, when scientists showed people could work out basic maths problems and read and analyse sentences without being consciously aware they were doing either. Their brains were finding answers to intelligence tests these people *did not even realize they had been asked.*

To investigate, the scientists used a technique called continuous flash suppression, in which pairs of special glasses present different images to each eye. Volunteers had their right eye bombarded with vivid and colourful and rapidly changing shapes. (They are called Mondrian patterns because they draw on the appearance of paintings by the Dutch abstract artist Piet Mondrian.) Meanwhile, their left eye was shown a series of simple arithmetic sums – such as what is eight plus seven plus three.

In these tests, the visual stimulus of the shapes to the right eye is so distracting it takes the conscious mind several seconds to even realize the left eye is shown something different. Before this could happen, the scientists took the images away. The volunteers had no conscious knowledge of the maths question at all. Yet, without them knowing the numbers were there, their subconscious was busy working out the answer.

After the images were turned off, the scientists flashed up a number to both eyes, and asked the volunteers to shout it out as quickly as possible. When the flashed number was the same as the answer to the sum – eighteen in the above example – the scientists found the volunteers said it significantly quicker, suggesting their subconscious had worked out the answer and primed them with the number.

The same thing happened with words. In a follow-up experiment, the scientists swapped the maths for a simple sentence. Some statements made sense, for example 'I made coffee', and some didn't, like 'I ironed coffee' and 'the window got mad at her'. This time the researchers left both stimuli running into both eyes, and asked the volunteers to say when the sentence – nonsense or sensible – popped into their head. They were looking for when the subconscious processing of the sentence produced a result flagged to the conscious mind.

The scientists thought the subconscious mind would call attention to the incongruous statements earlier, because the semantic violations they contained were surprising. They were right. These statements popped into the heads of the volunteers significantly faster. The scientists say this shows the subconscious minds of their volunteers were busy reading and working through the semantic meaning of the sentences, even while their conscious awareness was dominated by the colourful shapes. This would seem to support a popular explanation for savant skills: that they can dig into a type of subconscious processing denied to the rest of us. And it means that one way to seek neuroenhancement is to target these regions.

The principle that the unconscious mind can be triggered so it affects conscious activity is called subliminal priming, and it's pretty controversial. Subliminal advertising – flashing brief brand names and images – is banned in many countries and musicians from the Beatles to Judas Priest have been accused of planting hidden messages in songs.

Some subliminal priming seems little more than pranking and mischief making. If you can get hold of an old VHS copy of the Disney film *The Rescuers* and pause it around the 38-minute mark, when rodent heroes Bianca and Bernard fly past a building in a sardine box tied to the back of Orville (of Albatross Air Charter Service), then you can see – just – an image of a topless woman in one of the windows. The 1980s BBC show *The Young Ones* would regularly flash up random subliminal images – a frog in one episode and a skier in another – for no apparent reason, other than it seemed like a laugh and it would annoy people.

Subliminal priming annoys people in science too, mainly when they can't reproduce the findings of studies that claim, for example, that hearing words linked to the elderly, such as 'retirement', makes people walk, like old people, more slowly. Many studies in social psychology that have reported these kinds of effects are now under renewed scrutiny. There's no doubt subliminal priming can and does occur in specific circumstances; the question is how closely stimuli can be mapped onto specific behavioural responses given what else is going on both outside and inside the brain.

Scientists do agree that the subconscious mind can recognize patterns, even when the conscious mind is unaware of them. Some of the earliest experiments to show this were carried out at the University of Tulsa in the 1980s and 1990s. Volunteers, including a set of PhD students, were shown thousands of images that flashed briefly on a computer screen one after the other. They saw a simple grid, two squares wide and two squares high. In one of the four boxes was a letter or number and the volunteers simply had to press a key to indicate which – top left, top right, bottom left or bottom right. The target moved each time, seemingly at random.

However, the scientists had hidden an underlying rhythm in the sequence – the target moved around according to a set pattern. And unknown to the volunteers, their subconscious was learning it. The more they stared at the screen and pressed the buttons, the quicker and more accurate they got. Towards the end, the scientists threw down a googly and changed the imperceptible sequence. Sure enough, the responses slowed and more mistakes crept in.

Interviewed afterwards, not one of the volunteers was aware of the pattern they had learned and followed. Even more striking, they still couldn't find it when the scientists showed them the sequence of images as stills, and invited them to spot it. Their conscious mind could not match their unconscious ability to process the concrete information flooding into their eyes.

The same mechanism might explain the cocktail party effect – how you tune in immediately to a separate conversation when you hear your name mentioned. Your unconscious mind is constantly processing all the sounds it hears, but keeps them from your conscious awareness so you are not overwhelmed. Only when something is pertinent – like your name spoken by your boss half a room away – is it flagged for your attention.

Psychologists who have explored how these unconscious abilities vary between people say it does not seem to match differences in conscious intelligence – g, or proxies including IQ. People with higher IQs, in other words, aren't any better at unconsciously tracking and detecting patterns. This makes sense in evolutionary terms, they argue. Unconscious intelligence probably predates humans, and so relies on brain circuitry and regions different from conventional cognitive ability.

The way this older, deeper, subconscious brain can identify patterns and perform rudimentary calculations – and how some savants can access this – might combine to explain one of the most enduring and puzzling savant skills: calendar counting.

Calendars look ordered but are irregular, peculiar things. British and American calendars, for instance, miss out most of the first and second week of September 1752. At midnight on 2 September, those nations and others around the world skipped forward to 14 September. The dates that should have come and gone between never happened.

Not that Britain could complain. The loss of those eleven days was all its own doing. It hadn't been keen on the new Gregorian system of keeping track of time that many other European countries had long switched to because they agreed it was more accurate. Protestant Britain and its Empire, which relied on the older Julian system, refused to adopt what it regarded as a Catholic invention for as long as it could. By the time it admitted the new calendar was better, Britain had slipped so far behind that, when the switch did come, it had to skip forward eleven days.

Then there is the timing of Easter, which jumps around March and April seemingly at random, but is actually calculated according to an ancient formula that places it on the Sunday after the first full moon after the spring equinox. And leap years, of course, have an extra day because they introduce 29 February.

Calendar-counting savants can navigate all of this complexity. Among the best were a pair of twins in New York, George and Charles, who could name the day of any date some 40,000 years into the past or future. And, like most savants, they did not seem to consciously work out the answer. They said it just appeared in their head. It was a product of their subconscious processing.

Exactly how savant calendar calculators do it has puzzled psychologists for decades. From observing savants in action, and by timing them on their performance to identify days and dates from the near to the far future, they conclude it is a mixture of memory and calculation skills.

Calendars are peculiar but they have their own rhythms and patterns. There are only fourteen possible templates: 1 January can only fall on seven possible days and it might or might not be a leap year. The whole thing repeats itself every twenty-eight years, so the calendar for 2016 is the same as for 2044 and so on. If certain anchor points – Christmas Day in 2000 was a Monday – can be remembered, this provides a platform to work out the rest.

Mathematicians have produced various algorithms to mimic the calendar-counting skill of savants. One was

Lewis Carroll, author of *Alice's Adventures in Wonderland*, which itself contains many maths references and in-jokes. Another is John Conway, perhaps best known for inventing what is known as Conway's 'Game of Life' – a simple simulation of evolution and development called a cellular automaton, which spawned several generations of life simulation games, such as 'SimCity' and the rest.

In theory, most people could learn to use these anchor points and calculations to identify days from dates, at least for a span of a few decades. It takes time to work out the answer this way though – much longer than savants. It also demands plenty of conscious attention, and so it helps to have lots of conventional intelligence. Even among autistic savants, there tends to be a link between the speed and accuracy of their calendar calculation and their IQ.

In the 1960s, a psychologist in Oklahoma who visited George and Charles in New York became determined to work out how they and other calendar-counting savants do it. After much study of calendars he developed his own method and taught it, step-by-step, to one of his brightest postgraduate students. He then told him to go away and practise. The student was called Benjamin Langdon, and he made slow progress at first. Even with multiplication tables provided to ease the mental workload, for the first eight sessions he struggled to work out the correct day. After sixteen sessions, to Langdon's own amazement, his subconscious intelligence clicked into gear.

A decade later, one of his colleagues recalled:

Despite prodigious practice on Langdon's part, he could not match the speed of the twins' operation for a long time. Suddenly, he discovered he could in fact match the twins in speed. Somehow, surprising to Langdon, his brain had automated the complex calculations, had absorbed the table to be memorised with such effectiveness that now . . . he no longer had to consciously go through the various operations.

As his skill developed, so did the student's reluctance to discuss it, which defeated the purpose of the project, which was to test and report the method. Another colleague said:

> One interesting observation that I have always remembered was that when Benj became proficient and was giving rapid answers, he became irritated when he was pressed to tell us how he was doing it. The answer was there and it was not a step by step process for him. The process resembled what we now call implicit memory in that the right answer would be given but could not be explained by specific, explicit memories.

Implicit memory is another term used to describe the smart unconscious – it's the memory of procedures and habits which are stored and recalled by the brain without us realizing it. We discussed it earlier in the chapter on sports technique.

While many savants, including George and Charles, spend a lot of time examining calendars, they say they do not deliberately memorize them. And they are unlikely to have consciously worked out the algorithms the mathematicians found to identify the days. Their subconscious instead seems to harvest the relationships between the dates and days from the concrete raw details that passes in and out of their sensory memory. More importantly, they have access to the results of this subconscious processing. They can read it off.

If Benjamin Langdon managed to turn his calendar-calculating method into a subconscious task and in doing so developed a savant-like skill, this seems to support the idea other savants make use of these subconscious processes too. Further, it adds to the evidence that every brain has the subconscious capacity to develop savant skills of its own.

THIRTEEN

The Happiest Man on Death Row

As part of my experiment in cognitive enhancement, I went to meet my fellow Mensans to see what we had in common. To see what kind of people were united by high IQ scores, and to see if they came across as more intelligent or different in any other way.

Every year, Mensa invites its members to meet for a weekend of fun and games in a different British city. In September 2015, it was in Glasgow. Of the 300 or so attendees, I was the last to arrive. I was lucky to get there at all: I forgot to register before the deadline, got the dates mixed-up and then booked a hotel in the wrong part of the city.

The meeting ran from Thursday to Monday, but I could only make the Saturday night. So by the time I rocked up in the early evening at a new-looking hotel built across the Clyde from the steel armadillo that is the Glasgow Science Centre, the fun and games had been under way for some time. I know this because the first Mensa signs I followed led me to an actual Games Room. The used coffee cups and scattered empty chairs indicated the fun had finished, and seemed to have been replaced by two men arguing over how to use the computer.

Tables in the next room bulged with stapled paper. There were dozens and dozens of printed newsletters, with titles like 'Cognito', 'Now', 'Parnassus', 'Pathseeker' and 'Econo-mania', each produced by one of Mensa's special interest groups. What do people in Mensa do? Turns out that many of them write articles, edit newsletters and circulate them to other members who share an interest in country music, aircraft, cars, football, home recording studios, Americana, empathy, beekeeping, fridge magnet collections and at least two full tables-worth of more topics. I scooped up as many newsletters as I could carry and walked to the bar.

That's where I met John and Mary, who had both been members of Mensa since the 1970s. 'I started coming to these events to meet intelligent women,' John told me. While he obviously had, his appeal as an intelligent man had not brought the desired reward and he was still single. Mary told me: 'Most people are very friendly but watch out for the oddballs.' Then she hurriedly reassured me. The oddballs were very welcome too, and nobody in Mensa judged them.

Both said they kept their Mensa membership a secret from friends, and came to events like the Glasgow meeting because they were social and fun. They saw the same people each year and enjoyed catching-up, but mostly they did not see each other outside Mensa events. Why would they? We don't necessarily have anything in common just because we all have high IQs, Mary said.

As we talked about the society, John mentioned there had been a big Mensa scandal in the 1990s. I looked it up later. In 1995, the organization's long-standing chief executive Harold Gale was fired for running a private business from Mensa HQ in Wolverhampton. A tribunal and an internal enquiry followed and within a couple of years Gale was dead in a car accident. Before his death, Gale would tell people how he was ejected from the Wolverhampton offices in a 'dawn raid' led by then Mensa chairman, Sir Clive Sinclair, who then had the locks on the building changed.

After Sir Clive stepped down, Julie Baxter (IQ of 154) was appointed chair in 1997. She left after just nine months, complaining the Mensa committee was obsessed with 'self-aggrandisement and the pursuit of power for its own sake' and that some of the men on it were 'sad people with no social life' fixated on the organization. Things in Mensa were much calmer now, John said.

John asked me if I had joined any of the special interest groups. He edited one of the newsletters and told me about the pile on offer in the other room. I was about to reply that I had already found them when he added they were for reference only and I shouldn't take any away – Mensa didn't want non-members to read them. When John wasn't looking, I asked the barman for a bag to hide my stash.

The more people I talked to at the Glasgow meeting, the more I realized that nobody wanted to talk about what I was interested in: their IQ, how high it was, or how they thought it made them any different. One man said he had put his Mensa-endorsed IQ score on his CV, and then taken it off again. They didn't know each other's IQs and claimed they didn't want to. At first I thought people were embarrassed, but then I started to believe that they just didn't care. It was a social group, like any other, virtually everyone said, like joining a tennis club or a local history society.

Except of course, it wasn't. Tennis club members join to play the sport. Members of a local history society don't come along to discuss gardening or golf, they are united by a shared interest in local history, and that's what they want to talk about when they meet up. Yet, IQ and intelligence, the only thing that all of these Mensa members – many of whom had come to Glasgow on their own – knew they had in common with the person sitting next to them, seemed taboo.

The person I was sitting next to was called Charles and he ran the Mensa supervised tests to screen would-be members in one region of the country. That was lucky for me, as I was keen to gather some intelligence on how I might return and sit the test again. Journalists who sit the tests should be asked to declare themselves, he said, so I had been lucky there too. Charles said he was always sure to ask at each test session if a journalist was present, and to warn they couldn't interview any of the other people there. He repeated the rules, he said, when he did the PowerPoint presentation before each set of questions. I liked Charles but I was glad he hadn't run the test session I went to.

So if they didn't want to explore and compare IQs, why did these people join Mensa? Most I spoke with said they were simply curious, or that it had been more popular when they first became members in the 1970s. Britain was a kinder place then, many said, and less willing to ridicule people who were proud of their talents and abilities. A couple of them said they'd had bad experiences at school, either bullied or told they wouldn't achieve anything, and wanted to prove their worth. A couple of people said that, but I got the sense others felt the same, but were unwilling to tell me. A suspicious number of their 'friends' and 'people they knew' had joined for that reason, they said.

If joining Mensa had proved anything, even to themselves, I was not sure that it had helped them feel more at ease. When I got up to leave, the conversation around the table was about the glossy Mensa monthly magazine posted to members. The organization sent them in clear plastic bags. Some members were unhappy and wanted to receive them in unmarked brown envelopes. 'I don't want the postman to see them,' one man said, 'because he just takes the piss.'

As I walked out, I saw a group of people who had arrived to use the hotel gym look at my Mensa name badge, and at my plastic bag full of leaflets and special interest newsletters. The bag, I realized, said on the outside that it was for my dirty laundry with the hotel's compliments. I remembered what Mary had said about the oddballs and left them all to their games.

* * *

Do you know your IQ? Most people who say they do get it wrong. Friends tell me their IQ was measured in school as 160, 180, even as high as 200. That's possible, but it's extremely unlikely their IQ remains that high today. Their intelligence hasn't necessarily changed, but the way we measure it has.

IQ scores based on mental age tests, as Alfred Binet intended, are indeed a good way to identify children who might need extra attention. But they carry an obvious flaw. Simply getting older is enough to shrink IQ.*

That's why childhood IQ measured in this way is redundant in adulthood. My thirty-year-old friend who says she has an IQ of 160, for example, is probably referring to her ten-year-old self who was told she had a mental age of sixteen. For her IQ to be measured as 160 in that way today, she would have to possess the mental age of a forty-eight-year-old, which somehow doesn't sound as impressive, and is anyway absurd, because it assumes average performance on intelligence tests marches ever upwards into old age and beyond.

Using the ratio of mental and physical ages to calculate IQ is useless beyond the classroom, and is a source of continuing confusion, but it does have the merit of linguistic accuracy. The other way used to measure IQ – and the one celebrated by Mensa – does not even get the name right.

---

* If a five-year-old with an IQ of 200 sat the same IQ test a day later, which happened to be her sixth birthday, and scored a (very advanced) mental age of 10 again, her IQ would still shrink to 167 overnight.

The tests of intelligence used by Mensa do not present the results as a quotient. Instead, they use a statistical formula to compare an individual's performance to an average. The further the deviation from the average, the higher (or lower) the score awarded.

What's important here is this average does not reflect a real score on a real test. It is a number plucked from the air to *represent* average performance. The number used for IQ tests is 100, but that does not mean, as some newspaper stories about kids who get into Mensa imply, that someone with an IQ of 100 has answered 100 of 150 questions right, or achieved 100 per cent on something. The 100 is a score *awarded* for an average performance. It doesn't have to be 100, it could be set instead at 200 or 900, in the same way as football, rugby and tennis all award a different number of points for a goal, try and point.

A better-than-average performance on an IQ test gets someone a score higher than 100, but how much higher? It depends on the IQ scale used and is another reason why a single number – 'I have an IQ of 145' – is useless on its own. (Just as it's important to know if the sport is tennis or football when someone says the score is 15–0.)

The most commonly used IQ scoring system assumes two-thirds of people will have IQs of between 85 and 115. IQs below and above those boundaries get progressively rarer, until the distribution says only about one in fifty people will have an IQ below 70 or above 130. Towards the edges, the numbers drop off quickly. Only about one in 1600 has an IQ above 150, and only one in 30,000 above 160.

That's ultimately what modern IQ tests measure. Rather than comparing your performance to your age, they compare it to everybody else's performance. This is important – IQs are relative. However the overall performance of a group may change, there will always be individuals within that group with IQs at the low and high end of the scale. It's impossible for everyone to have an IQ of above 100, no matter how much we educate, selectively breed or cognitively enhance. When people say things like, 'nearly half of Americans have an IQ of under 100' as a criticism, they reveal more about their own intelligence than anyone else's.

We have a curious relationship with intelligence these days. Rather than looking down on people with lower IQs, as was common when the feeble-minded were ridiculed, much public scorn is reserved today for those towards the upper end of the scale. Perhaps this is down to envy and jealousy, as the benefits of mental ability become more pronounced, or maybe it's a reflection of a society that has fallen out of love with expertise. The admonishment is especially severe for those who both choose to sit a Mensa test and then pay to join. Browse any online forum when the subject of high-IQ clubs comes up and the comments are almost as derogatory as some of those made about the feeble-minded a century ago.

'I qualify for Mensa, but upon looking into who they are, I realized it's just a club for the socially inept, because more often than not intellects [*sic*] somehow wind up socially retarded.'

And: 'Highly intelligent people are usually incredibly stupid.'

And: 'The few people I've known who joined Mensa were misfits who (or whose parents) wanted to try and compensate for some deep sense of insecurity and inadequacy by having something they could think was bigger than "normal" people's'.

The modern relationship is most awkward when it comes to high intelligence in children. Some surveys suggest as many as one in five girls and one in ten boys at secondary schools hide their ability at maths, chiefly to avoid being picked on and bullied.

Then there are the prodigies. A prodigy is a bit different to a savant – while a savant does things beyond most of us, it is the age at which a prodigy achieves things, rather than their fantastic nature, which draws attention. Few prodigies drew as much attention as William Sidis. Plenty of people study for a maths degree at Harvard University; William Sidis became famous because he did so when he was eleven. By seventeen, he was teaching undergraduates at what is now Rice University in Houston, and a year later he returned to Harvard to take a second degree, this time in law.

For a while, Sidis was described as America's most famous child. When he arrived at Harvard, newspaper reporters asked his opinion on national politics. When he suggested a radium-powered rocket could reach Venus in twenty minutes, the *Chicago Tribune* made it front page news. And when the newly arrived Sidis delivered a Harvard lecture to staff at the maths department, the *New York Times* compared it to the boyhood preaching of Jesus Christ. (This was a full fifty-six years before the same paper changed musical history with a front page story about an Alabama radio DJ who refused to play Beatles' records because John Lennon had said in a months-old interview they were more popular than Jesus.)

Alongside the plaudits for Sidis, others in the media liked to kick sand in his face. Reporters mocked his lack of interest in sport and pointed out he was afraid of dogs. A vowed celibate, when Sidis said in an interview it was awkward when girls flirted with him, the response was ferocious. 'A man can't learn about women from books, especially calculus books,' one newspaper scolded. Asked for her opinion on the fuss by a New York newspaper, a twenty-year-old woman from Dallas who had never met Sidis said, 'I bet that he is a sissy, sports a wrist watch and wears his handkerchief in his sleeve.'

Sidis is known as much today for what he failed to achieve in adult life. Halfway through his law studies, he was arrested when a socialist May Day parade turned violent. To avoid prison, his parents took him to California, after which Sidis sought a quiet life. He took a series of low-grade and menial jobs. This was, he said, so he did not have to use his mind.

The press gloried in his apparent fall. Even the great minds of the US legal system joined in. After Sidis sued the *New*

Yorker magazine for breach of privacy over an article that mocked him, the court of appeals concluded: 'Even if Sidis had loathed public attention we think his uncommon achievements and personality would have made the attention permissible.' It added: 'His subsequent history, containing as it did an answer to the question of whether or not he had fulfilled his early promise, was still a matter for public concern.'

Another much more recent prodigy who landed with a bump is Sufiah Yusof, who went to Oxford University to begin a maths degree in 1997 aged just thirteen. Two years later, she disappeared, and was eventually found after a high-profile police search working as a waitress in a cafe. In March 2008, a tabloid newspaper revealed she had later become a prostitute.

The tone of the coverage of such cases is almost gloating, as if these young prodigies somehow made claims with their early high achievement that they could not justify; as if their unusual intelligence was a deliberate ploy to annoy the rest of us. Of all the sins of youth, cognitive precociousness seems one of the hardest to forgive.

Sometimes, Mensa and its members do themselves no favours when it comes to the organization's public image. In the week before Christmas 2012, the BBC and Mensa member and spokesman Peter Baimbridge were both forced to make grovelling apologies after an interview live on *BBC Breakfast* in which Baimbridge described people with IQs of 60 and below as root vegetables. 'So most IQ tests will have Mr and Mrs Average scoring 100, and the higher you get, the brighter you are. And if your IQ is somewhere around 60 then you are probably a carrot,' Baimbridge said.

Amid the (deserved) criticism he received, many of the responses featured a common belief about those such as Baimbridge with a high IQ. Some of you are probably thinking it now. It's neatly summarized by liamf12 of Oxford, who wrote in the online comment section beneath a news story about the carrot apology in the web version of the *Daily Mail*: 'That's the problem in this country, firms are now fast tracking graduates into management roles when my experience of intellectuals is that they can give the right answer to a question but have a distinct lack of common sense.'

After the sister of European Space Agency scientist Matt Taylor – one of a team who landed a spacecraft on a speeding comet in 2014 – told reporters her brother sometimes struggled to remember where he parked his car, the *Daily Telegraph* was moved to ask its readers the same question: why do geniuses lack common sense?

Common sense is the claimed kryptonite of the super-intelligent, the Achilles heel of having a good IQ score. Common sense, in fact, dictates that the more intelligence someone has, the less common sense they can fit in their oversized brain. Unlike IQ, which is shown to correlate positively with life success, albeit indicated by some pretty crude metrics, common sense is believed to not just fail to show the same link, but to actually decrease with IQ. That's a pretty bold assumption, and an assumption seems to be as far as it goes. There is no hard evidence that IQ is negatively associated with common sense; mainly because common sense is just as hard to define as intelligence.

Common sense is typically described as a kind of practical intelligence.* It's usually measured as a judgement on someone's decision making, but the verdict on whether someone shows common sense or not seems to come down to whether or not the person doing the judging agrees with the particular decision made. As we saw, high IQ tends (on average) towards left-leaning liberalism, so perhaps it's not surprising one of the most common targets for failing to show a lack of common sense is the (left-leaning liberal) idea of political correctness.

In 2009, a professor of theoretical medicine called Bruce Charlton wrote an academic exploration of the apparent high IQ–low common sense paradox, which he dubbed the clever sillies. High intelligence was an asset when humans were evolving, he said, because it enabled abstract analysis to tackle serious problems in the distant past that could be life-or-death situations. But modern humans, he continued, don't need this trait as much. In fact, we have by now developed mature and reliable specific responses to most of the situations we find ourselves in, including social encounters. He says these domain-specific shared responses are what most people mean by common sense – the generally accepted and road-tested way to do things.

---

* Unlike presumably the impractical kind that gave us the internet, heroic decreases in childhood mortality and the A380 super-jumbo.

High IQ people, Charlton suggests, are different. They simply can't resist the temptation to continue to deploy their abstract problem-solving skills in even familiar situations, for which the best options have already been approved by the rest of the community. They are driven to find novel solutions, at the expense of the tried and trusted common sense. And many of these ideas are wrong, or worse, ridiculous.

It's an interesting – if speculative – idea. But it would seem to succeed or fail based on the strength of the examples: what kind of novel solutions are wrongly offered by the clever sillies to what kind of established problems? And, frustratingly, Charlton doesn't offer any. He talks about physical scientists being silly outside work and social scientists being silly both in and outside of work. Abstract analysis of social problems, he suggests, produces left-wing political views. He bundles the work of the clever-sillies together as 'political correctness' in which, 'foolish and false ideas have become moralistically-enforced among the ruling intellectual elite. And these ideas have invaded academic, political and social discourse.'*

A dislike of political correctness is today a common stamp of conservatism, so perhaps deciding who is a clever silly just comes down to one's politics. Or maybe

---

* Dr Charlton wrote a book *Thought Prison* about how the rise of political correctness meant that Western civilization was doomed. Again, he didn't specify examples – asking readers to supply their own from personal experience and unofficial knowledge.

the desire to attribute a lack of common sense to intelligent people is just another version of the scorn they receive from some quarters, which may itself be a reaction to the historical superiority claimed by those esteemed men of the Mutual Autopsy Society and their eugenic friends. It's certainly an impression that some go out of their way to promote.

Albert Einstein is often said to have had little common sense. The evidence? He didn't like to wear socks and he sometimes got lost when walking around Princeton (where he lived well into his seventies) and had to ask for directions to his house. This was a man, it barely needs to be said, who was acutely aware of the impact of his own decision making, and in one case – urging US President Roosevelt to build an atomic bomb – was tortured by the consequences for years afterwards. It is hard to argue with the view of the man himself. 'Common sense,' Einstein said, 'is the collection of prejudices acquired by age eighteen.'

While the nature of common sense and intelligence may be elusive, the concept of g – general intelligence, as identified by Charles Spearman and accepted by most psychologists – is easier to grasp. It's mental prowess, a measure of how well someone can turn their brain and their talents to a number of different tasks, all of which involve some degree of cognitive processing. Emotional intelligence, managerial intelligence, sexual intelligence and the rest – if they employ the higher functions of the brain then they almost certainly all rely on the same foundation of g.

And despite its bad press and the way new interpretations of intelligence try to set themselves up as superior, logic would suggest that, on average, someone's IQ score reflects their general intelligence. Plenty of psychologists are happy to accept the link, and in fact they pose a challenge to critics, to those who insist IQ is not a reliable measure of g. Find something else, they demand. Conceive a separate measure of cognitive ability. And then show this new measure, while reliably indicating general intelligence, is independent of IQ.

It can't be done, or at least it hasn't been done so far. Simply put, IQ *is* a pretty good indication of intelligence, or at least the best way we have found so far to try to constrain and quantify intelligence.

But it's also true that IQ does not – and cannot – cover the entire spectrum of human abilities that go into judging if someone is intelligent. Or indeed, whether they lack intelligence. Some of the sharpest and most controversial arguments over IQ are over the way it is applied at the bottom end of the scale. While Alfred Binet devised the original intelligence tests because he wanted to identify and help low-achievers, for some of these people the concept has become a way to kill them. In many American states, someone's IQ, and how it relates to their intelligence, is quite literally a matter of life and death.

On death row, the lights burn all night. Joe Arridy didn't mind that, as it gave him more time to play with his toy train. It was a wind-up train with two carriages. Sometimes,

Joe would reach through the bars of his cell and send the train puffing down the jailhouse corridor. Then he would squeal with excitement when one of the other men sent the toy back to him – a prison guard or one of his fellow condemned prisoners in a neighbouring cage. Usually it was Norman Wharton, who was there waiting to die because he killed a policeman. Wharton would play with Joe and his train in this way for hours. Back and forth. Choo-choo!

When they tired of watching the game, wife-killer Angelo Agnes and Pete Catalana, murderer and drug-dealer, would reach out from their own cells and tip the train onto its side. 'A wreck! A wreck!' Joe would shout with glee. 'Fix the wreck!'

Joe Arridy was twenty-three but he had a mental age closer to a toddler. Roy Best, the warden at Colorado State Prison in Canon City, gave Joe the train as a present for the Christmas of 1938, Joe's last. Less than two weeks later, Best and Father Schaller, the prison chaplain, came to Joe's cell, to accompany him on the short walk up the gravel road to the death house on Woodpecker Hill. Joe wanted to take his train with him, but Agnes said he would hold it.

Best and Schaller had tried to explain to Joe what was about to happen, but he didn't understand. He was grinning when he walked into the gas chamber. He was still smiling when, wearing just a pair of white shorts and socks, he sat on the middle of the offered three chairs and allowed the guards to fit the straps. Only when the black bandage was placed over his eyes did Joe sense something was not right, and then only for a moment. The grin returned when

Best squeezed his hand on his way out. Schaller, tears in his eyes, said goodbye and turned to leave too. The steel door closed.

Joe Arridy was gassed because he had confessed to the rape and murder of a fifteen-year-old girl called Dorothy Drain. It was a miscarriage of justice. No physical evidence put him at the scene, yet the arresting sheriff claimed Joe accurately described, and in great detail, the pattern of wallpaper on the victim's bedroom wall, her clothes, and his own complex feelings of remorse for the crime.

In fact, Joe could not distinguish red from black or name the days of the week. When Father Schaller read the Lord's Prayer with him shortly before his death, he had to do so two words at a time so Joe could keep up. Joe was innocent of the crime for which he was condemned, but that is not the point. He was not fit to stand trial in the first place.

Of the 3,100 prisoners sentenced to death and held in US prisons, a fifth are believed to have some degree of intellectual impairment. There are two ways off death row for these people – in a coffin, or to convince an appeal judge they should be spared because they are mentally retarded, their intelligence is low enough to have their sentence commuted to life imprisonment.

The US forbids as cruel and unusual the execution of people with abnormally low mental ability and several states have long used IQ tests to decide the intelligence of convicts and so their fate; the so-called Bright Lines approach. Florida set the bar at an IQ of 70; a score of 71 or above is enough to see someone sentenced to die, while

Oklahoma used an IQ threshold of 75. (The Supreme Court intervened in 2014 and told states they needed to be more flexible, but the principle of a cut-off remains.)

Even scoring below the cut-off point has not always been enough to spare people the death penalty. Prosecutors have argued, successfully in some cases, that black and Latin American defendants should have a few points added to their measured IQ scores, to take into account the relatively poor performance of those ethnic groups on intelligence tests. These defendants, the legal argument goes, score low because of social and cultural factors, not because they are intellectually disabled. When their scores are 'adjusted' to allow for this, the legal system can execute them. To save people on death row, defence lawyers have to show their clients are not smart enough to die. And to do that, they call Steve Greenspan.

Steve Greenspan is a scientist who has spent his life wrestling the notion of intelligence. In his day job he works as an educational psychologist at the University of Connecticut, studying neuropsychological skills such as reason and judgement, and how they can go wrong. In his spare time he tries to save the lives of condemned prisoners.

Greenspan, for instance, told a Louisiana court in November 2013 a killer called Teddy Chester was mentally retarded and so should not be executed for the murder of a taxi driver in 1995. Chester resented what he saw as an insult, and in a catch-22 situation, argued he was not mentally disabled at all. He said witnesses who described him as a slow learner were lying. Well, his defence said, that is what someone with difficulty in reasoning *would* say. As I finished this book in late 2016, Chester remains scheduled for execution.

In 2009, Steve Greenspan became involved in work to secure a posthumous pardon for Joe Arridy, a campaign that began not as they often do with a new piece of evidence or a deathbed confession from the real killer, who had been tried and executed years ago. It started with a poem.

Called 'The Clinic', by Marguerite Young, the poem was published in 1944. Just twenty lines long, it describes the scene in the Canon City prison the day of Joe's execution: a weeping warden, a toy train and the killing of a dry-eyed child.

In 1992, the poem found its way to a Rocky Mountains man called Robert Perske. Perske had previously worked as a chaplain at an institute for people with intellectual disabilities, assisting them when they found themselves on the wrong side of both the law and the legal system. When he read the poem, he thought he might still have it in him to help one more, and resolved to find the dry-eyed child who was forced to leave behind his train.

His search took him through every local history and library archive in a string of cities on the eastern side of the Colorado Rockies. He pieced the details together from scanned boxes of microfilms that preserved the reports of long-gone local newspapers: *The Pueblo Chieftain*, *Grand Junction Daily Sentinel*, *Wyoming State Tribune* and the *Rocky Mountain News*. By 1995 he had Joe's name, and a decade later a website dedicated to the campaign to clear his name.

As part of that campaign, Steve Greenspan wrote a detailed report on Joe's low intelligence, based on records uncovered by the people working to clear his name. Greenspan tore apart any claim Joe Arridy could have understood the crime he was accused of, the concept of execution, or the evidence presented at the trial – or that he was able to judge right from wrong. Joe Arridy, he said, had a mental age of four and a half.

Nothing suggested Joe Arridy killed Dorothy Drain; there were no witnesses and no signs he was even near her house. Greenspan's evidence showed he was unable to have even given the confession that saw him gassed. He was a classic patsy – impressionable, uncomprehending and passive – for police more interested in pursuing their own prejudiced agenda than solving crimes.

Psychiatrists at his trial said Joe was mentally deficient. Joe's lawyer said this deficiency meant he should be found not guilty due to insanity – he could not, for example, tell right from wrong. But the psychiatrists also told the court Joe was not insane: a person had to be normal before he could lose sanity, and, they said, 'this defendant has never been normal'. A confused jury sent him to his death.

Joe Arridy never understood he was to be killed or what it meant. Roy Best, the warden who was ordered to execute him, said Joe was the happiest man who ever lived on death row. He spent his final days in his cell making faces in a polished dinner plate. On 5 January 1938, Best asked what he wanted for a last meal.

'Ice cream,' said Joe.*

A number of scientists, including Steve Greenspan, are trying to get the legal system to broaden its view of intelligence beyond IQ, to take into account the many different ways it can be defined, and so the equally many ways a lack of intelligence can be identified.

In the latest *Diagnostic and Statistical Manual of Mental Disorders*, the American Psychiatric Association scrapped IQ scores as the primary way to diagnose intellectual disability (previously called mental retardation). Instead it emphasizes the impact of cognitive ability on behaviour. Greenspan and other psychologists are trying to persuade the US courts to view intelligence through the same broader lens. They are developing tests of how someone can conceive abstract ideas such as numbers and time, their self-esteem and ability to follow laws, and practical skills including use of money and efforts to protect their own health.

In 2013, the first such test was released, called the Diagnostic Adaptive Behaviour Scale. Just like IQ, it too boils down the concept of intelligence to a single number – which means it could be used by the courts to distinguish people. The benchmark score on the test is 100; someone with a score of 70 or below would be judged to have an

---

* Joe was officially pardoned by the governor of Colorado in 2011.

intellectual disability. For someone on death row, that could be enough to save their life, in a way an IQ test, which measures a much narrower set of criteria, would not.

In the search for intelligence tests beyond IQ, Greenspan has examined the role of gullibility: how likely someone is to go along with the suggestion of others and not think through the consequences. This can be framed as a failure to use intelligence wisely under certain social conditions, and even people who we think of as highly intelligent can behave in a gullible way.

In November 2012, Oxford University-educated physics professor Paul Frampton was convicted and sentenced to fifty-six months for drug smuggling in Argentina. He claimed to have been the victim of a scam after supposedly meeting a model on a dating website, and said he had been tricked by gangsters into transporting cocaine hidden in the lining of a suitcase. A common reaction to such cases is to ask how someone could be so, not gullible, but *stupid*. Greenspan thinks such people might have a cognitive style relying too heavily on intuition, and this makes it hard for them to turn their large intellects to certain situations. They are book-smart, not street-smart.

Gullibility can make people with learning disabilities vulnerable to sexual abuse and practical jokes, but it can also make otherwise intelligent people do extremely foolish things, including castrate themselves and commit suicide. In 1997, some thirty-eight members of the Heaven's Gate cult in California were found dead after they killed themselves because they were told a spaceship was parked behind a passing comet and was going to pick up their souls. Some victims made farewell videos, some of which are on the internet, and they appear bright and articulate people. Indeed, Greenspan and his colleagues have searched in vain for any sign people with mental retardation are more likely to be seduced by the messages and promises of such groups.

A pattern emerges. While people with intellectual disability tend to be gullible and so act foolishly in social situations, allowing themselves to be duped and coerced by those who seek to take advantage, the stupid actions of the otherwise intelligent tend to be in practical domains and entered into voluntarily. Stories of people who found ingenious ways to accidentally hurt or kill themselves were a staple of the early internet years, and were codified as the Darwin Awards – awarded to people who remove their genes from the collective pool in an 'extraordinarily idiotic manner'.

This could be down to different personality types. Impulsive risk-takers are more likely to crash and burn than those who rarely stray from solid ground. Or it could be conventional ways to think about intelligence miss out a crucial ability – to think and behave rationally. The psychologist Keith Stanovich at the University of Toronto has coined the term 'dysrationalia' to describe people who struggle with rationality, just as people with dyslexia have specific difficulty with language. Rational thinking, he says, should be measured and called RQ. And just as high intelligence is no protection from dyslexia, so a top IQ score would be no guarantee the same person will have a high RQ.

To think rationally is to act according to one's goals and beliefs. But it is also to form and hold beliefs supported by available evidence. Humans are generally accepted as the only animal capable of rational thought, and one reason we are considered to be the most intelligent. But humans are also the only animals who can think *irrationally*. We all have a number of cognitive biases that try to pull us away from reason. One of the most important is the myside or confirmation bias – the way people gather and assess evidence tends to be in line with their existing opinions. Another is the way we jump to conclusions and decisions, making as little cognitive effort as possible. The most famous example is: A bat and ball cost £1.10 in total. The bat costs £1 more than the ball. How much is the ball? Most people quickly say (or at least think) 10p but that's wrong (the total then would be £1.20).

This thinking style and the difficulty of avoiding it (and even its potential power) have been discussed at length for years. What matters here to the question of intelligence is that susceptibility to these types of irrational thought does not seem to be strongly linked to IQ.

Because of this, and also because, as we've seen, theories of intelligence tend to prosper when they are presented as counter to IQ, Stanovich argues strongly his idea of rational thought is a separate cognitive category. Intelligence is IQ, he says, and rationality is different to both.

Yet, as rational thought is measured in part by the actions it directs, rationality seems to be covered by the catch-all broad definition of intelligence being something you use to get what you want. To act according to beliefs to achieve goals, however irrational those beliefs might be, demands intelligence.

If Stanovich is wrong and intelligence and rationality do overlap, this is good news for our efforts to enhance intelligence. For rationality can be increased. Rational thought can be encouraged by making people aware of the types of cognitive bias that constantly seek to undermine it. No clever neuroscience necessary. Just thinking about circumstances in a different way, reframing the question, can usually help.

Example: a guaranteed way to double your chance of winning on the national lottery exists. Want to know what it is? Buy a second ticket with different numbers. Congratulations, your chances have soared from one in fourteen million to one in seven million. Don't order the sports car just yet.

That sounds facile, but it's the principle relied on by everyone from con artists to advertisers to persuade people to part with their money. It's why headlines about environmental risks 'doubling' the chance of cancer scare people with no good reason, and why politicians and special interest groups can so successfully hide misleading agendas behind a phrase to seduce our cognitive biases. How gullible are you? Send £50 for our searching questionnaire.

FOURTEEN

On the Brain Train

A year after my first test, I decided to use a combination of neuroenhancement methods to cheat my way back into Mensa – to use smart drugs and brain stimulation together for the maximum effect. The bad news with this strategy was that, should my IQ increase, I wouldn't be able to say with any certainty whether it was the modafinil or the electrical stimulation that gave the most help. But I would hopefully get a useful answer on the effectiveness of neuroenhancement – on my brain at least.

I had plenty of modafinil pills left so that side was easy: I would get up early on the day of the test and take one with breakfast. That should give the drug three to four hours to work its way into my system. The electrical brain stimulation was more of a challenge. I toyed with the idea of wearing my Spiderman hat to the test to conceal the electrodes, and flicking the switch to light up my brain at the start of each set of questions. That seemed a risky strategy, partly because it would be pretty obvious what I was up to, and partly because it would mean nipping out now and then to rewet the sponges.

Not all brain stimulation studies need the electric current to flow when the desired task is being completed. Many of the psychiatric experiments stimulate the brain separately to the therapy sessions, for example, to prepare it or soften it up. They do it every day, or every other day, as part of the treatment routine. That seemed a decent compromise, and would at least save me from sitting my Mensa retest in a strange hat while warm salty water dripped down my face. In the week leading up to the test, I would use my internet-bought kit to stimulate my brain at home for twenty minutes each day.

But which part of the brain to target? Certainly not the motor cortex I had aimed for in my rowing experiment, as I wasn't seeking improvement in the muscles and other bits of the body it controls. Better to stimulate a part of the brain more involved with cognition and processing. I found studies that seemed the most useful and applicable to the Mensa test. They were carried out by a neuroscientist called Allan Snyder at the University of Sydney in Australia.

Snyder divides opinions among other scientists, and not just because he likes to wear a curious cap at a jaunty angle in all of his television interviews. He calls his brain stimulator a 'thinking cap' and is not shy about making strong claims for what he believes it can do, including boosting human creativity.

In a series of tests, Snyder has used low-frequency magnetic stimulation to inhibit the left anterior temporal lobe, to try to release dormant savant skills he argues are present in everybody's brain. And he has obtained some interesting results. After his brain stimulation, several volunteers were better able to proof-read and find the mistake in the following text:

A bird in the
the hand is worth
two in the bush

Drawing skills also improved, with a similar change in emphasis, from abstract to concrete, to the dementia patients who suffered damage to the same site in the brain. Sketches were judged as more life-like and complex.

Some people also showed an improvement in numerosity: a savant skill that enables rapid counting of a large number of objects. It's immortalized in *Rain Man* when Raymond counts cocktail sticks from a box dropped by a waitress. In *The Man Who Mistook His Wife for a Hat*, Oliver Sacks reported that George and Charles – the New York calendar-counting twins – could do it as well. (He changed their names in the book.)

A box of matches on their table fell, and discharged its contents on the floor: '111,' they both cried simultaneously; and then, in a murmur, John said '37'. Michael repeated this, John said it a third time and stopped. I counted the matches – it took me some time – and there were 111. 'How could you count the matches so quickly?' I asked.

'We didn't count,' they said. 'We saw the 111.' Similar tales are told of Zacharias Dase, the number prodigy, who would instantly call out '183' or '79' if a pile of peas was poured out, and indicate as best he could – he was also a dullard – that he did not count the peas, but just 'saw' their number, as a whole, in a flash.

'And why did you murmur "37", and repeat it three times?' I asked the twins. They said in unison, '37, 37, 37, 111'.

In Snyder's magnet experiments, a computer flashed up repeated images of between fifty and 150 dots for a second and a half, and people were asked to guess how many appeared each time. Ten of the twelve volunteers improved after stimulation.

In 2012, Snyder turned to electrical stimulation, and this time he used the different electrodes to both inhibit the left side and activate the right side of the volunteer brains. He reported a similar improvement – this time in the ability of people to solve a classic puzzle that requires a bit of lateral thinking. It's called the nine dot problem, and it asks simply that you connect all nine dots using just four straight lines, without lifting your pen up or retracing a line.

Try it:

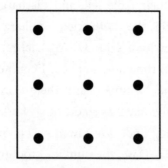

(SPOILER ALERT. The solution is at the top of the next page.)

The only way to solve the puzzle is to extend the lines beyond the boundary of the square. Like so:

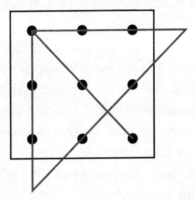

In Snyder's tests, only the people who had the electrical brain stimulation realized that.

Well, almost. One volunteer who turned up at the lab, a man called Brian in his early twenties, was not allowed to participate, because in conversation it emerged that he had suffered a significant head injury as a child. Still, he was interested in the research and, as he was already there, asked to have a go at the puzzle.

He solved it – the only volunteer to do so without stimulation to inhibit the left anterior temporal lobe. After he did, Brian told the scientists how he saw the world in concrete terms:

I only focus on a particular thing, so if I walk into a room, I'd just take things methodically, each thing at the time, I don't look at the whole picture . . . I notice everything by itself, as singular objects instead of the whole scene . . .

283

even my writing . . . I'm only focused on one part . . . My long term memory is very very good . . . I can recall everything that happened in year 6 [when he was twelve years old].

Brian offered to dig out his medical records. His neurologist had noted that Brian had suffered multiple injuries to the left hemisphere, and a 'fracture at the left temporal bone' – the same spot the experiments were trying to turn down by simulating damage.

So, I would copy this study and aim my brain stimulator at the anterior temporal lobes.

My home-grown stimulation of the anterior temporal lobes began with a bang. I wet the electrodes and tucked them under the hat, one on each side of my head. And then as I turned the switch from off to 2mA an extraordinary thing happened. A flash of light shot across my vision, a tracer bullet passed through my brain. That hadn't happened previously when I targeted the motor cortex. I gasped, and my wife – already nervous about my self-experimentation – leaned forward ready to pull the plug on it.

It was a phosphene, a pinprick of light not there and created only by the electrical stimulation of the retinas at the back of my eyes, or more likely my brain's visual cortex. It's a well-known effect and similar to the patterns that swim in your head should you press your eyes for too long, or the sprinkle of stars that appear when you stand too quickly. Phosphenes are thought to be responsible for the phenomena of Prisoner's Cinema, in which people confined to darkness for a long time experience what they describe as a light show.

Phosphenes are harmless, but they do show that messing with the brain using electricity can have unexpected effects. At the moment, there is no evidence of harmful side effects, but electrical brain stimulation can certainly go wrong. This seems more to do with poorly assembled and used devices, rather than an inherent hazard with the technology itself, but some users have said they burnt themselves quite badly.

As such, I wouldn't want to be seen in this book as *encouraging* people to try it for themselves, but then, I don't need to. The reports of success, both in the scientific and consumer media, are doing that already. And as the technology is developed for medical and other uses, then demand will surely rise with awareness. Companies are selling the kit and volunteers are queuing to buy it. Some scientists have warned against DIY brain stimulation, and others say sale and use should be regulated. It's a debate that has some way to run, but we can't pretend that self-neuroenhancement is not happening. Better, in my view, to explore its power and limitations and gather the information we need to make an informed decision. And the public, surely, must play a role in that. If brain stimulation and other neuroenhancement technology does work, then plenty of them are going to want to use it. And that's whether or not it's regulated, controlled and risk-free.

My Mensa retest was at the same London university as

the first, this time on a muggy Saturday morning the day after it was announced that Britain voted to leave the European Union. The rejection of expertise – and by extension, intelligence – had played a major role in the referendum. A string of smart people, from Stephen Hawking to the head of the Bank of England, had warned of the wide-ranging negative consequences. 'I think people in this country have had enough of experts,' responded Michael Gove, the newspaper-journalist-turned-politician who was chief among those pushing Britain to go it alone.

As a member, I didn't think that Mensa would have allowed me to take the test again. So I assumed the identity of my brother for the day, I gave the Mensa man in charge some convincing ID and he found his name on the list.

We were relegated to the basement this time. There were more candidates too – twenty including me – with about the same mix of ages and, as far as I could tell, nationalities. The invigilator was from the same school of instruction and supervision as Charles, who I had met at the Glasgow meeting. He didn't ask journalists to declare themselves, but he did explain the procedure in long and repetitive detail. He even drew multiple choice boxes on the white board to show us how to fill them in, and how to correct mistakes using the rubbers on the ends of the pencils he gave to each of us. Maybe he just thought we weren't very bright.

The questions were the same, more or less. Certainly on the second of the two papers, the language test, I recognized some. On the first it was harder to know – one sequence of interlocked circles and triangles tends to look the same as the next.

I had taken the modafinil after breakfast and by mid-morning when the test began, the drug was moving through the gears in my brain. The sense of alert and focused concentration it had given me the first time was back. I whizzed through the early easy questions, but as time ticked on and the puzzles got tricky, a curious thing happened. The modafinil – at least I think it was the modafinil – dragged me fully into each question, and made it more difficult to take an educated guess and move on. Where I could see the answer early on, the drug acted as an accelerator. But when some effort was required, it was almost a brake. I was sucked into the problem, the way it was phrased and posed, and, if I was taking too long, I found it harder to walk away from the intellectual challenge and move on to the next question.

For example, one question towards the end described the circumstances around the fate of a lost explorer, and asked us to work out whether it was thirst, cannibals or prowling lions that did for him in the end. It was a full paragraph of text and a test of logic, memory and reason and was worth just a single mark, the same as dozens of previous questions, but I found that I couldn't move on from it. I read each word and as I did so, I saw the man's plight, recorded in the final desperate pages of his diary. He was hungry, thirsty, scared – hunted. What happened to him? The possible sequences of events played out in my head and each time I needed to get to the climax (always the same for the unfortunate explorer) before I could consider the next. I was lost in the abstract. I was wasting valuable time considering his motives, his fear of wild animals and how I would react in that situation. I couldn't easily extract and work with the concrete information, the facts and the sequence of events.

(In January 2017 scientists in Germany reported what looks like a similar effect of modafinil on expert chess players. Those given the drug made better moves, but actually lost more games on time penalties because they took so long to choose them.)

In the end, I'm pretty sure I got the question about the dead explorer right (the key was to consider the threat the lions posed to the cannibals) but as I filled out the relevant box to record my victory, the man from Mensa said time was up, and I had left four questions unanswered. I was still thinking about the lions as he whisked the answer paper off my desk, and did not even think this time to tick box A for the rest.

The results, he said, would be posted within a week.

Smart drugs and electric brain stimulators are the current focus of research in neuroenhancement but that isn't where the story begins and ends. A few years before scientists started doping students and playing with currents, they looked to classical music as a way to increase intelligence.

When General Augusto Pinochet established a military dictatorship in Chile, following a violent coup in 1973, political opponents turned to an unlikely weapon of resistance. Crowds would gather outside detention centres and sing a familiar song. The words were in Spanish. The tune, unmistakably, was Beethoven's 'Ode to Joy'.

Considered by many to be the composer's masterpiece, 'Ode to Joy' features at the climax of his ninth symphony. The lyrics are taken from a German poem, with lines that celebrate brotherhood and the unity of mankind. As such, it has also been played at protests to signal disapproval with other totalitarian regimes, such as in apartheid South Africa and by the students who squatted in Tiananmen Square.

On 14 January 1998, an American politician called Zell Miller played 'Ode to Joy' at the Georgia state house on a portable tape recorder, as part of annual budget negotiations. 'Now,' he asked his colleagues, 'don't you feel smarter already?'

Miller, state governor, wanted them to approve his plan to spend taxpayer dollars on CDs and cassettes of classical music, which he would then give to new parents to play to their babies. Tens of thousands of infants each year, he argued, would have their brains stimulated and their intelligence enhanced. Miller grew up in the mountains of north Georgia, where he had witnessed for himself the intellectual benefits of music. 'Musicians were folks that not only could play a fiddle but they were also good mechanics. They could fix your car.'

Not to be outsmarted, the neighbouring state of Florida passed its own law to harness the same effect. State-funded day-care centres, the law insisted, must play classical music to the children for at least one hour every day.

Florida and Georgia were acting on a scientific study that suggested, yes, hearing classical music could boost intelligence. A 1993 experiment got college students to listen to verbal relaxation exercises, or nothing, or to the first movement 'Allegro con Spirito' from Mozart's Sonata for Two Pianos in D Major, and then asked each group to answer some simple questions to test spatial awareness. The students who listened to Mozart experienced a temporary, but significant, increase in scores.

Christened the Mozart effect, the results of the study made headlines around the world. But in doing so, the conclusions and implications came unglued from the actual results. Nobody is sure how, but a limited study of a narrow cognitive test in young adults morphed into the promise of a general lift in intelligence for children and babies. And why just Mozart? Surely any classical music would do, right?

Even as the politicians were pouncing on the study results, other scientists were starting to pick holes in them. Most importantly, despite dozens of attempts, other scientific labs couldn't get Mozart's music to boost the test scores of their own volunteers in the same way. The idea still has some defenders, but the attitude of most researchers to this twenty-year-old controversy can be summed up by the title of a 2010 summary and analysis of all the various trials published in the journal *Intelligence*: 'Mozart effect – Shmozart effect'.

Equal disagreement surrounds another widely acclaimed method that promises to boost and protect cognitive ability. So-called brain training – a series of repetitive tasks usually done on a computer – is now a multibillion pound industry, but every now and then a group of esteemed neuroscientists will pipe up to say people who sign up for it are wasting their time and money.

The disagreement is usually not over whether brain training tasks can have an effect, but whether or not that effect is transferable to everyday activities. One task, for example, gets people to remember and repeat endless strings of numbers for several hours each week. The evidence is pretty strong that, yes, these people do improve their ability to remember and repeat strings of numbers. But it's much less clear whether that benefit will endure and, more importantly, if these people could apply it to something more useful – remembering to pick up a pint of milk on their way back from work perhaps.

In one of the largest efforts to test the benefits of brain training, scientists asked viewers of the BBC science television show *Bang Goes the Theory* to try it for themselves as part of a rigorous experiment. More than 50,000 people registered for the tests, and 11,430 went as far as carrying out the minimum two online training sessions a week for the requested six weeks.

As well as the strings of numbers to memorize, some of the volunteers were trained in a test of grammatical reasoning, which is known to reflect someone's general intelligence. They were shown pictures and as quickly as possible had to say whether simple statements were true or false: for example, 'the circle is larger than the square'. Another task addressed spatial memory; finding and remembering the location of stars hidden in boxes. And a related game asked volunteers to remember what objects (hat, ball, etc) were concealed in a series of shuttered windows. Each game was geared so that, the more correct responses, the harder and faster it got.

Practice, it seems, does (help) make perfect. After six weeks, plenty of people were scoring higher on the specific test they had been asked to train on. But only on those. When the tasks were switched, those who had been trained at finding stars showed no significant improvement at finding hats (all volunteers had tried each of the different tests at the start so the scientists had something to compare). This was true even when the different tests should have tested the same cognitive function, such as abstract reasoning.

This is a problem for the popular claim that these cognitive tests somehow 'train' the brain and can increase 'mental fitness'. While increased physical fitness, from regular running for example, would be expected to help people better perform other exercises, like cycling, the benefit from the brain training was less useful, and less versatile. Bang goes the theory.

Things get slightly more positive when the volunteers are grouped by age. If cognition and intelligence can be improved then the most important societal use of such a tool would probably not be to improve the thinking skills of the young and healthy, but to prevent their loss in the elderly. As more people live well into their seventies and eighties, dementia is becoming a huge burden in developed countries and soaking up an increasing share of healthcare costs. Indeed, some brain training games are explicitly marketed as one way for older players to stave off the ravages of degenerative conditions like Alzheimer's disease.

When treated as a separate group, the thousands of silver-haired viewers of *Bang Goes the Theory* who trained every day on online tests did seem to show some useful improvements in mental ability. Volunteers aged sixty and above scored higher on a scale used to assess independence. Called the instrumental activities of daily living scale, it measures how well people say they can get around and do stuff – using the telephone, shopping, cooking, washing clothes, keeping on top of personal finances and other tasks that people start to struggle with as their mental faculties decline with age.

It's pushing it a little to say this study backs up the clinical claims that brain training could ease the impacts of dementia, but it does suggest something that deserves more investigation.

Despite the often trenchant criticism made of brain training and the lack of evidence for useful and transferable effects, there doesn't seem to be a massive downside, even if the benefit is marginal. These tests are games after all, and people pay to play games all the time. Hype of brain training seems different to the dangerous claims that are made for some untested medicines, because those are too often presented as an alternative to evidence-backed main-stream therapies.

At present, there are no reliable techniques to try to avoid the misery of Alzheimer's, so nothing to distract people from. If the brain training games don't work then that's a shame – and yes, some people have become rich off the back of us trying – but it seems a reasonable and sensible position for people to have a go and see. It takes years to accumulate the kind of evidence that will convince the most robust of sceptics and some of the people who are trying to train their brains, frankly, don't have years.

Besides, the sceptics who argue it's all a waste of time have a natural human failing on their side: laziness. Brain training is a huge effort. Based on the (almost zero) amount of transferable improvement the younger people showed in the simple tasks of memory, scientists estimated it would take four years of brain training for someone to be able to remember a single extra digit. The over-sixties who showed the benefit above only did so after they fiddled with the online tasks for at least ten minutes a day, every day for six months.

All studies like this lose a huge proportion of their volunteers as they proceed because people just can't be bothered to keep it up. It's the same with physical training as well, as anyone who has pledged to do simple stuff like a few dozen sit-ups or ten minutes' brisk walking a day knows full well. Just like the real thing, plenty of people pay to join these brain gyms, try it for a couple of weeks and then fail to trouble the equipment again.

The sad truth is that as it stands, most of us will see most of our cognitive functions decline and fall as we age. Vocabulary and general knowledge – classic types of crystallized intelligence – are robust and can keep gradually increasing well past someone's seventieth birthday. But even without disease and dementia, fluid intelligence, reasoning and problem-solving usually rises quickly with schooling and teenage development and peaks in early adulthood.

For pre-schoolers, don't be fooled into paying for DVDs that promise to make babies smarter. In 2009, Disney was forced to offer refunds to millions of parents who bought 'Baby Einstein' videos – containing music, puppets, and bright colours – because the company had claimed they were educational. They're not. But then maybe we don't need to go in search of cognitive enhancement, for ourselves or our children. Maybe we can just sit back and let cognitive enhancement come to us.

Old houses with their lower door frames are tough on modern visitors. Doors were made shorter in those days because the people were shorter – figures show that in developed countries people are now a full four inches higher on average than they were 150 years ago.

This creep upwards poses problems beyond not being able to fit in old houses – modern footballers find it harder to kick the ball past the now-giant goalkeepers that block their path. In 1996, officials at the global governing body of the sport, FIFA, even floated the idea that the size of the goalposts should be increased, to create more empty space for the strikers to shoot at.

Nobody is sure why we are getting taller. And here's the striking thing – it's not just height that is increasing in this way. So is intelligence. Across the developed world, each generation consistently scores better on IQ tests. Even as we focus on cognitive skills, argue about the best way to educate, and struggle to define intelligence, the ground is shifting beneath our feet. For decades, children have been consistently turning out smarter than their parents and grandparents. The dream of the eugenicists for a future race of superior humans is being realized. Millions, perhaps billions of people, are being cognitively enhanced. Something is shifting the goalposts. But what?

The steady rise in IQ across the citizens of developed nations is called the Flynn effect, after the New Zealand political scientist James Flynn, who was among the first to report it. Flynn had noticed a curious thing: people seemed to find older IQ tests easier. The questions looked the same, but a volunteer's IQ measured on a test given in 1940, say, was significantly higher than if they took a newer test, published in 1980. Because IQ is measured relative to the average score of a population, and the person taking both tests was the same, the change in score could mean only one thing: the average score of the population in 1940 and 1980 was different. Specifically, the average score for the 1940 test must have been significantly lower to allow the same ability to appear relatively superior on that test.

It works the other way around, too. Get different groups

to sit the same test, or compare the number of right answers from groups who took the same test in the past, and the younger generation always does better. Significantly better – the average Flynn effect is about three full IQ points a decade. So people born in Britain in 1990 have, on average, IQ scores a massive fifteen points higher than the generation born during the Second World War. The US saw the average IQ rise by fourteen points from 1932 to 1978, and Japan witnessed a nineteen point rise between 1940 and 1965. All of these people, and the societies they lived in and contributed to, were cognitively enhanced by birthday. The change is visible on other forms of intelligence test too.

There are some tentative signs that these IQ increases are manifesting themselves in the real world. The average age of the top international chess players declined from about thirty-five to twenty-five over the last decades of the twentieth century. Scientific productivity, measured by research papers and patents published, has increased massively. And the number of children in US schools diagnosed as suffering from mild mental retardation has seen a decrease. None of this is conclusive of course – these things could all be changing due to reasons other than increasing average IQ – but it does appear to be in line with what we might expect.

Not everybody accepts that society is getting smarter. A hard core of intelligence researchers is reluctant to give up the concerns of the eugenicists that the modern world is doomed to idiocy. Despite the IQ increases observed in many places, this group makes the case that there has in fact been a decline in general intelligence since the Victorian days. They even use the results of all those reaction time measurements conducted by Francis Galton on the prime minister and the rest to try to prove it. Average reaction times in western countries, they say, are slower these days, which shows that less intelligent people (some of whom are immigrants) have been having too many kids and spoiling it for everyone else. Per capita rates of 'innovation and genius appear to have declined subsequently'.

If anything, it's the opposite conclusion that deserves serious consideration. If IQ in Britain and the US and other industrialized countries has been rising steadily, then when did that rise begin? And just how low was average intelligence before? Clearly there were always some high achievers who caught the eye (and built bridges, cracked trigonometry, predicted celestial mechanics, wrote the US constitution, invented the bicycle) but were our great-grandparents and people further back generally, well, just a bit thick?

Intelligence, recall, is using what you've got to do what you want. Or to get what you need. And 150 years ago, before remote controls and tube lines and having to go to school and having to work in the knowledge economy and having other people want to know what you know and how much, people wanted and needed different things from their brains. James Flynn has looked at all the records he can find from as many places as kept them and suggests that, prior to industrialization, humans focused on concrete objects and as modernity shaped their lives, so their brains learned to grapple with abstract concepts. That type of abstract thinking, the memory and visualization and spatial awareness and the ability to make connections beyond the surface, are the bits of intelligence that an IQ test tends to pick up.

Flynn describes it like this:

> The industrial revolution demands a better educated work force, not just to fill new elite positions but to upgrade the average working person, progressing from literacy to grade school to high school to university. Women enter the work force. Better standards of living nourish better brains. Family size drops so that adults dominate the home's vocabulary and modern parenting develops (encouraging the child's potential for education). People's professions exercise their minds rather than asking for physically-demanding repetitive work. Leisure at least allows cognitively demanding activity rather than mere recuperation from work. The world's new visual environment develops so that abstract images dominate our minds and we can 'picture' the world and its possibilities rather than merely describe it.

This process gives nations as they develop about 100–150 years of IQ gains, until both the social and intellectual transformation levels off. At some point, universal education is established and family size is as low as it will go. Leisure time becomes saturated with hobbies.

National IQ scores do suggest that some countries have

reached this plateau phase, and their populations have hit the brain-power ceiling. The Scandinavian countries in particular seem to have compressed the development cycle and peaked, perhaps aided by heavy state investment in education and welfare. Countries in East Asia such as Japan and Korea, alongside Britain, Germany and the US, are approaching the plateau, while late starters such as Brazil, Argentina, Kenya and Turkey are just hitting the sweet spot and witnessing massive IQ gains. Other low income and developing countries have yet to get going, which can per-haps explain why average IQ measured in those places is typically on the low side.

Flynn says:

Modernity means breaking from simply manipulating the concrete world for use. It means classifying, using logic on the abstract, pictorial reasoning and more vocabulary. The IQ test items that have risen over time make the same cog-nitive demands. The enormous score gains are a symptom of the radically new habits of mind that distinguish us from our immediate ancestors.

Exactly how modern life raises IQ remains undetermined. There are plenty of suggestions, which range from kids and adults getting wise to how to sit and pass tests – guessing at multiple choice questions when time is running out, for example – and better nutrition, maternal and pre-natal care, to larger brains inside the larger heads on all those bigger bodies. Some have even suggested the brighter minds are down to brighter houses and the spread of artificial light.

None of these seem as likely as the cognitive impact of education, and how it has developed, and the teaching of mathematics especially. Sophisticated geometry, algebra and multi-step problems have replaced rote learning, and are presented earlier in the curriculum to younger and younger kids. A 2005 analysis by education experts concluded: 'Currently, young children regularly engage in visual-spatial problem-solving associated with prefrontal based working memory functions their grandparents' generation would not have been exposed to until [aged eleven and twelve] and their great-grandparents' generation may not have been introduced to at all.'

Whatever the exact cause of the Flynn effect, its IQ rise is almost certainly not down to genetics. A few generations within a century or so is too short a time and too large a change to be down to nature – especially as modern studies of the genetics of intelligence show the impact of genes on g and IQ is much more subtle than a couple of pieces of DNA to be altered. Still, that does not mean that deliberate modification of genes – genetic engineering of humans – could not raise IQ even further.

One problem with this futuristic idea is that several massive studies of intelligence have failed to find any specific genes responsible for IQ differences, at least in the normal and high range (many genes have been identified that seem to play an important role in mental retardation). That doesn't mean the genes are not there. It means they are not obvious, which means there are probably an awful lot of them, each of which has a small effect on its own.

While many geneticists might see this huge number of genes that direct intelligence as an obstacle to tinkering with DNA to boost IQ, Stephen Hsu views it as an opportunity. Hsu is a physicist at Michigan State University and he likes to think big. In 2014, he wrote an article for the online magazine *Nautilus* as big as it comes. It was titled: 'Super-intelligent humans are coming. Genetic engineering will one day create the smartest humans who have ever lived.'

Confusingly for those planning to pop by, in 2007 the Beijing Genomics Institute moved its headquarters to Shenzhen, some 1,300 miles from the Chinese capital. Bill Gates visited a few years later and left dumbstruck by what he had seen. The building hummed with electronics, but not the kind that made Gates his fortune. The machines he saw were crunching human capital. They were analysing, measuring, sequencing and recording the secrets of DNA sourced from thousands of people.

The institute goes by the less geographically misleading (if more opaque) corporate abbreviation BGI these days, but its purpose remains much the same – to work out what, genetically, makes us tick. And in 2012, news started to leak from Shenzhen that the company was turning its DNA sequencers to investigate the genetics of intelligence. It was asking for high achievers, including top scientists and people with high scores on IQ tests, to volunteer a sample for analysis. One of its advisers is Stephen Hsu.

If the individual genes (or the version of the genes) with a (minimal) positive effect on IQ can be identified, Hsu says, the precise genetic requirements of the super-intelligent human could be mapped – and produced in a living embryo. Natural genes could be changed. Each of the 'off' bulbs in the string of lights could be twisted to 'on'. And the result would be dazzling.

In his article, Hsu claimed: 'Given that there are many thousands of potential positive variants, the implication is clear: if a person could be engineered to have the positive version of each causal variant, they might exhibit cognitive ability which is roughly 100 standard deviations above average. This corresponds to more than 1,000 IQ points.'

An IQ of over a thousand? 'It is not at all clear that IQ scores have any meaning in this range,' Hsu admitted.

> However, we can be confident that, whatever it means, ability of this kind would far exceed the maximum ability among the approximately 100 billion total individuals who have ever lived. We can imagine savant-like capabilities that, in a maximal type, might be present all at once: nearly perfect recall of images and language; super-fast thinking and calculation; powerful geometric visualization, even in higher dimensions, the ability to execute multiple analyses or trains of thought in parallel at the same time; the list goes on.

That would be the ideal, but, Hsu continued, plenty of intellectual gains could still be obtained with more modest genetic tweaks. Switching 100 gene variants from 'off' to 'on', for example, could produce a gain of 15 IQ points – easily the difference between a child struggling at school and sailing through college.

It sounds far-fetched, but then the future usually does. The first part of Hsu's plan, the identification of the thousands of genetic variants implicated in intelligence, is largely a numbers game. We know the genetic influence is there. Throw enough DNA from enough smart people through enough sequencing machines, and build computers powerful enough to analyse the data that spews from the other end, and they will probably be found – or at least enough of them to try to make a difference.

The editing of the genes into shape is more technically demanding and probably not possible yet, but progress is racing ahead. Even in the eighteen months or so I have been writing this book, the scientific world has been turned upside down by the rapid rise of a new gene editing technology called Crispr-Cas9. It allows scientists to make precise and accurate genetic modifications, and puts the ability to do so in the hands of even non-expert researchers. Chinese scientists stunned everybody in biology in 2015 when they announced they had already used the tool to modify the DNA of a human embryo*, but virtually all the work so far has been in animals.

A few months after the work with the human embryo,

---

* This was a research project only, and the embryo was never intended to be developed into a person.

scientists edited the embryonic DNA of Bama pigs – about half the size of regular farm pigs – to turn off genes involved in growth. The modified pig embryos, when implanted into a surrogate sow and grown to term, produced mini-pigs. These pinched porcines, when full grown, were just a sixth of the size of a farmyard pig – about the same as a dog. The scientists knew exactly what to do with them. They sold them as pets. The same team are now working to customize the colour and pattern of the pigs' coats, which they say they will then be able to customize on demand. The research was done, incidentally, at BGI Shenzhen.

FIFTEEN

Faster, Stronger, Smarter

The second envelope from Mensa was waiting for me when I returned from work, poking out beneath a gas bill. I opened the gas bill first. Its numbers were higher than I expected. I hoped the same would be true of the letter that announced my new IQ.

It was. My cognitively enhanced score on the language test had crept up to 156, from 154 before. And on the Culture Fair test, the tough one with the symbols, it had soared to 137, from 128. That put me on the ninety-ninth percentile on both.

My IQ as measured by the symbols test – the one I had tried to improve on using the brain stimulation – was now 135, up from 125, and so well above the threshold required for Mensa membership. In the year and three months since the first test, or perhaps just in the week I had been stimulating my brain, my intelligence measured by that method had increased enough to overtake 3 per cent of the adult population, more than a million people in the UK.

Was the increase down to the neuroenhancement? It's

impossible to know for sure, but I think at least some of it was. As we saw already, a retest effect with IQ should see scores increase second time around. How much would still linger after a year? Scientists don't know, though Mensa is happy for people to try to get in as many times as they like, as long as they wait a year between tests. (And they are, of course, also happy to take their money as they do so.)

Second time around, each question still takes working out and, although I suppose it's possible my subconscious remembered the individual puzzles and chipped in with some help, it certainly felt like I had to start from scratch each time. I was definitely more prepared, and I knew going in I would have to be quick – but I learned that pretty rapidly the first time around too.

Perhaps the increase was down to mere statistical chance? All measurements of human performance come with natural variability and are influenced by how much we've slept, what we've had to eat and drink, whether we are a morning or an afternoon person, the temperature of the room, whether the person at the next desk is coughing or tapping their pencil against their teeth and numerous other sources of influence, on both a conscious and sub-conscious level. Certainly, my relatively slight increase on the language-based test could easily be attributed to that. Such a small rise could even be down to pure luck and me guessing the answer as B rather than D a couple of times.

The larger increase on the Culture Fair test seems harder to entirely explain that way, though it's difficult to be too definitive. Most psychologists talk of IQ ranges and how confident they can be a score falls within a spread.

The most common of these is probably the 95 per cent confidence interval, which works out as an error margin of about five points both up and down. So, from my initial measured score of 125, we can conclude a 95 per cent chance my true IQ was between 120 and 130. And for the second, the measured IQ of 135, there is the same chance it is now between 130 and 140. It's not that simple (the spread tends to bulge towards the lower scores) and of course, there is a 5 per cent chance in both cases my actual IQ will fall outside the ten point spread.

It's worth remembering test scores – from IQ scores to exam grades – in the real world don't come with error bars. Most of us get a single shot at most opportunities to prove ourselves, and we have to live with the results. If statistical clouds of variation essentially make scores of 69 per cent and 71 per cent on a three-hour exam the same, well, nobody tells university examiners that. Score above 70 per cent on my undergraduate exams and you were awarded one degree classification and below 70 per cent it was another. To find a way to perform better on the day, to nudge that score from 69 per cent to 70 per cent, can have a massive impact on someone's life. And if it can be achieved with cognitive enhancement, then that means the technology and its effects are hugely significant for society.

To pick up easy money as a postgraduate I used to invigilate degree exams. On one boiling June day with pneumatic drills bashing away to dig up the road outside, I remember one distressed finalist putting his hand up halfway through a session and asking me, almost with tears in his eyes, if the stifling heat and noise would be taken into account when his paper was marked. Yes it would, I told him, knowing full well it wouldn't. Somehow, I think telling him that, statistically speaking, overlapping error bars on a high-scoring lower second-class degree and a low-scoring upper second-class degree meant the outcomes were essentially the same, wouldn't have reassured him much; especially not if a couple of dropped marks in that exam saw his life pivot on a lost opportunity.

There is always a cut-off point so people who fall on either side of it will be separated on some measure. That's not fair, but we all go along with it – given two candidates for a job with everything else equal, who would not choose the one with the higher grades?

On my IQ tests, maybe I just got lucky second time around and that could explain the higher score. Or maybe the practice, drugs and brain stimulation put me in a position where I could make the most of that luck. This is a world, remember, where until recently an IQ score of 70 would see someone executed and a score of 69 would let them live. Try telling that person a rise of a single IQ point carries no statistical difference.

Then there's the placebo effect. I knew I was taking genuine modafinil and I knew the current was flowing through the electrode sponges squeezed against my head. More, I knew they might increase my IQ, and, for the sake of having a decent ending for this book, I wanted them to.

Maybe I subconsciously tried harder in the second test (I don't *think* I did so deliberately) or maybe I had more confidence because I believed my efforts at cognitive enhancement had made me more intelligent.

It's hard to disentangle all of the confounding factors, which is why science and medicine don't take one-off results in such uncontrolled trials seriously as hard evidence. My experiment was not scientific and I generated no reliable data. Even if the cognitive enhancement effect is genuine we can't tell if one of the methods worked better than the other. I'm only a case study. But case studies can still be useful. They can identify effects that require attention, exploration and, eventually, explanation.

One explanation for the rise in my measured IQ is a combination of the retest and placebo effect with statistical chance. Another is the modafinil and brain stimulation between them had a genuine effect. (Another is Mensa simply marked either the first or second test incorrectly.) To find out, to explore and explain, the only way is to pay attention and to carry out larger and more controlled trials.

Should we? I think we should, if for no other reason than to give society the evidence it needs to decide what to do about cognitive enhancement.

Of all of the questions raised in this book, the medical and technical and neurological, the most important – and the most difficult to answer – seem to be those around ethics. Opinions on the impact of cognitive enhancement and the need for scrutiny and regulation, for example, will probably come down to how realistic and powerful we think the impacts on society could be.

At its most far-reaching, the stakes are huge. The impacts of the silicon chip revolution continue to claim more jobs each year: improved communications and automation have already hollowed out blue-collar jobs. Now technological progress is coming for careers of the middle classes; those for which school tests and exam grades are considered a reliable way to pick the most able – the most intelligent – applicants. The population is growing and opportunities are shrinking. Something will have to give. In that market, cognitive enhancement could be a vital and fought-over tool to help people get on.

Even if we don't get that far – if the mental phase transition remains a difficult and fragile effect to conjure on demand, and the promising results of experiments in the lab are hard to replicate in the real world – then the investigation and discussion about the principle are still useful. Intelligence has been on the scientific black list for too long. The topic deserves more than embarrassed looks and half-truths and the whiff of scandal when someone brings it up. If a new focus on the promise of techniques to increase intelligence forces broader contemplation of how we think about and relate to one of our oldest and most significant human abilities, then neuroenhancement and the neuroscience revolution will undoubtedly help more of us reach our full potential.

Charles Spearman's discovery of the positive manifold and his work on general intelligence created a great schism, and one that still runs through society like words through a stick of rock. How much is our mental ability given to us, and how much do we have to earn it?

Unearned privilege can be uncomfortable to associate with human value, for it carries too many reminders of the straitjacket of social stratification and the entitlement of the aristocracy. We prefer people to work for what they have, and expect rewards and status for those who do so.

(Ironically, it is often the people who believe intelligence is a natural gift, given more to some and less to others, who argue the most vociferously that it reveals something about the value of an individual.)

Cognitive enhancement offers a new twist on this century-old argument. If intelligence, in whatever form, is something people have to work for, if cognitive ability can be trained and improved and released with effort, then it's pretty simple to make the case that neuroenhancement undermines this effort and is cheating. If one person has access to a short-cut others do not, then the playing field is tilted in their favour.

Yet if intelligence is an immutable quality spread across the population, with some landing more in the heads of a fortunate minority, then the playing field is already biased against the rest. Why shouldn't those who lose out in the lottery of life have the chance to turn to technology to close the gap? Only when all have the identical chance and the baseline is levelled, can the performance of any human ability be truly said to reflect value, or more accurately, can the difference in performance be said to reflect higher or lower value.

I don't have the answers to all of these ethical questions. I don't know if anybody does. But I know the search for them will shape the way we allow and encourage cosmetic neuroscience to change society. They will set the boundaries and parameters of the world that we and our children and grandchildren live in. And, in all honesty, when it comes to the treatment of intelligence and the differences in intelligence, we can't do much worse than the generations of our parents and grandparents, who made a bit of a hash of it. We have an opportunity to do things in a better and fairer way, and cognitive enhancement – and discussions of cognitive enhancement – offers one tool to do that.

The drumbeat of the neuroscience revolution is growing louder. We should listen and prepare, consider and reflect on the choices available. We must acknowledge the clamour for change. And we must let the possibilities, risks and opportunities into our society on our own terms. Because, like it or not, they are coming in anyway – even if they have to smash down the gates.

ACKNOWLEDGEMENTS

I wasn't clever enough to write this book alone. So a big thank you must go to Jon Butler, Robin Harvie and especially Cindy Chan who helped me to organize the ideas, pages and words. Thanks also to my agent Karolina Sutton for her always firm and wise guidance. Stuart Ritchie and Alison Abbott were kind enough to read early drafts and offer some useful comments – thank you. And thanks to the good people at Mensa for making me feel so welcome at your meeting in Glasgow (and please forgive the subterfuge that followed). Numerous conversations with friends and colleagues contributed ideas – thanks all. Thanks to my wife Natalie for not freaking out when I told her I was going to electrify my brain, and for her continued support. Thanks to my daughter Lara for letting me include her picture, and to my son Dylan for not being too cross about me not including one of his. This book is for my parents, who this year celebrate their golden wedding anniversary – a true work of genius.

REFERENCES

INTRODUCTION

'rewiring', Mason L. *et al.* (2017), 'Brain connectivity changes occurring following cognitive behavioural therapy for psychosis predict long-term recovery', *Translational Psychiatry* 7, 17 January, e1001.

'pregnant woman', Sreeraj V. *et al.* (2016), 'Monotherapy with tDCS for treatment of depressive episode during pregnancy: a case report', *Brain Stimulation* 9 (3), pp. 457–458.

'catatonic schizophrenia', Shiozawa P. *et al.* (2013), 'Transcranial direct current stimulation for catatonic schizophrenia: A case study', *Schizophrenia Research* 146, pp. 374–375.

ONE: Our Brain Revolution

'charcoal', Anonymous (1890), 'The First Execution by Electrocution in Electric Chair', *New York Herald*, 7 August.

'expert panel', Hyman S. *et al.* (2013), 'Pharmacological cognitive enhancement in healthy people: potential and concerns', *Neuropharmacology* 64, pp. 8–12.

'oxygen chambers', Yu R. *et al.* (2015), 'Cognitive enhancement of healthy young adults with hyperbaric oxygen: a preliminary resting state fMRI study', *Clinical Neurophysiology* 126, pp. 2058–2067.

'briefing note', POST (2007), 'Better Brains', 285 June.

'old joke', I *think* this is from Billy Connolly.

TWO: Mensa Material

'less intelligent kids', Binet, A. (1905), 'Le problème des enfants anormaux', *Revue des Revues* 54, pp. 308–325. Translated in Nicolas S. *et al.*

(2013), 'Sick? Or slow? On the origins of intelligence as a psychological object', *Intelligence* 41, pp. 699–711.

'the scale', Sullivan W. (1912), 'Feeble-mindedness and the measurement of the intelligence by the method of Binet and Simon', *The Lancet*, 23 March, pp. 777–780.

THREE: A Problem of Intelligence

'workplace', for example, Kuncel N. and Hezlett S. (2010), 'Fact and Fiction in Cognitive Ability Testing for Admissions and Hiring Decisions', *Current Directions in Psychological Science* 19 (6), pp. 339–345.

'there's more', Ritchie S. (2015), *Intelligence: All That Matters* (John Murray Learning), pp. 40–54.

'Robert Jordan', AP (1999), 'Judge Rules that Police Can Bar High I.Q. Scores', *New York Times*, 9 September.

'1921', Anonymous (1921), 'Intelligence and Its Measurement: A Symposium', *Journal of Educational Psychology* 12 (3), pp. 123–147.

'follow-up', Sternberg R. and Detterman D. (Eds.) (1986), *What is Intelligence? Contemporary Viewpoints on Its Nature and Definition* (Norwood).

'Switzerland', Legg S. and Hutter M. (2007), 'A Collection of Definitions of Intelligence', *arXiv:0706.3639v1*, 25 June.

'stupid behaviour', Aczel B. *et al.* (2015), 'What is stupid? People's conception of unintelligent behaviour', *Intelligence* 53, pp. 51–58.

'Asian peasants', Luria A. (1976), *Cognitive Development: Its Cultural and Social Foundations* (Harvard University Press).

'Galton', see Holt J. (2005), 'Measure for Measure. The Strange Science of Francis Galton', *New Yorker*, 24 January.

'Spearman', Spearman C. (1904), 'General intelligence objectively determined and measured', *American Journal of Psychology* 15, pp. 201–293.

FOUR: Treating and Cheating

'review of studies', Kekic M. *et al.* (2016), 'A systematic review of the clinical efficacy of transcranial direct current stimulation in psychiatric disorders', *Journal of Psychiatric Research* 74, pp. 70–86.

'special needs school', Geddes L. (2015), 'Brain Stimulation in Children Spurs Hope – and Concern', *Nature*, 23 September.

'improve the scores', Chua E. *et al.* (2017), 'Effects of HD-tDCS on memory and metamemory for general knowledge questions that vary by difficulty', *Brain Stimulation* 10 (2), pp. 231–241.

FIVE: Pills and Skills

'Japan was heavily into meth', Alexander J. (2013), 'Japan's *hiropon* panic: resident non-Japanese and the 1950s meth crisis', *International Journal of Drug Policy* 24, pp. 238–243.

'UK officials', Boseley S. (2014), '£200,000 Smart Drugs Seizure Prompts Alarm Over Rising UK Sales', *Guardian*, 24 October.

'sharpen reaction times', see Franke A. and Bagusat C. (2015), 'Use of caffeine for cognitive enhancement', *Coffee in Health and Disease Prevention* (Elsevier), pp. 721–727.

'neuroscientists think', for example, Minzenberg M. and Carter C. (2008), 'Modafinil: A Review of Neurochemical Actions and Effects on Cognition', *Neuropsychopharmacology* 33, pp. 1477–1502.

'Harvard and Oxford', Battleday R. and Brem A. (2015), 'Modafinil for cognitive neuroenhancement in healthy non-sleep-deprived subjects: a systematic review', *European Neuropsychopharmacology* 25 (11), pp. 1865–1881.

'dramatically relapsed', Tan O. *et al.* (2008), 'Exacerbation of obsessions with modafinil in two patients with medication-responsive OCD', *Primary Care Companion of the Journal of Clinical Psychiatry* 10 (2), pp. 164–165.

'hypersexuality', Bulut S. *et al.* (2015), 'Hypersexuality after modafinil treatment: A case report', *Journal of Pharmacy and Pharmacology* 3, pp. 39–41.

'amateur athletes', Dietz P. *et al.* (2013), 'Associations between physical and cognitive doping – a cross-sectional study in 2,997 triathletes', *PLOS One* 11 (8), pp. 1–10.

'team managers', Rodenberg R. and Holden T. (2016), 'Cognition enhancing drugs ('nootropics'): time to include coaches and team executives in doping tests?', *British Journal of Sports Medicine*, January 25.

SIX: The Mutual Autopsy Society

'Whitman', Burrell B. (2003), 'The Strange Fate of Whitman's Brain', *Walt Whitman Quarterly Review* 20 (3), pp. 107–133.

'Mutual Autopsy Society', Anonymous (1889), 'A "Mutual Autopsy Society"', *The Lancet*, 19 October, p. 809.

'**skull collections**', Juzda E. (2009), 'Skulls, science and the spoils of war: craniological studies at the United States Army Medical Museum, 1868–1900', *Studies in the History and Philosophy of Biological and Biomedical Sciences* 40, pp. 156–167.

'**league tables**', Spitzka E. (1902), 'Contributions to the encephalic anatomy of the races. First paper – three Eskimo brains from Smiths sound', *American Journal of Anatomy* 2 (1), pp. 25–71.

'**Russian poet**', Vein A. and Matt-Schieman M. (2008), 'Famous Russian brains: historical attempts to understand intelligence', *Brain* 131, pp. 583–590.

'**Czolgosz**', MacDonald C. and Spitzka E. (1902), 'The trial, execution, necropsy and mental status of Leon F. Czolgosz', *The Lancet*, 8 February, pp. 352–356.

'**disclaim it**', Spitzka E. (1901), 'Report of autopsy on assassin disclaimed', *JAMA* 19, p. 1262.

'**father died**', Unknown (1914), 'Dr Spitzka's Brain Weighs 1,400 Grams', *New York Times*, 15 January.

'**Robert the Bruce**', Deary I. *et al.* (2007), 'Skull size and intelligence, and King Robert Bruce's IQ', *Intelligence* 35, pp. 519–525.

'**Albert Einstein's Brain**', for example, Hines T. (2014), 'Neuromythology of Einstein's brain', *Brain and Cognition* 88, pp. 21–25.

'**P-FIT**', Jung R. and Haler R. (2007), 'The Parieto-Frontal Integration Theory (P-FIT) of intelligence: converging neuroimaging evidence', *Behavioural and Brain Sciences* 30, pp. 135–187.

'**P300**' Amin H. *et al.* (2015), 'P300 correlates with learning and memory abilities and fluid intelligence', *Journal of NeuroEngineering and Rehabilitation*, 12 (1)87.

'**highly personal**', Finn E. *et al.* (2015), 'Functional connectome fingerprinting: identifying individuals using patterns of brain connectivity', *Nature Neuroscience* 18 (11), pp. 1664–1673.

'**as bad as it sounds**', Anonymous (1926), 'Racial Purification', *Nature*, 20 February, pp. 257–259.

SEVEN: Born with Brains

'**no longer teach**', Hunt E. (2014), 'Teaching intelligence: why, why it is hard and perhaps how to do it', *Intelligence* 42, pp. 156–165.

'**alarm bells**', Parens E. and Appelbaum P. (2015), 'An introduction to thinking about trustworthy research into the genetics of intelligence', *The Genetics of Intelligence*, Hastings Centre report 45 (5), pp. 2–8.

'**segregate**', Radford J. (1991), 'Sterilization versus segregation: control of the feebleminded, 1900–1938', *Social Science Medicine* 33 (4), p. 449–458.

'**nymphomaniacs**', Philo C. (1997), 'Across the water: reviewing geographic studies of asylums and other mental health facilities', *Health and Place* 3 (2), pp. 73-89.

'**Wedgwood**', Woodhouse J. (1982), 'Eugenics and the feeble-minded: the Parliamentary debates of 1912–14', *Journal of the History of Education Society* 11 (2), pp. 127–137.

'**young boy**', Butterworth J. (1911), 'The diagnosis of feeble-mindedness in school children', *Public Health*, August, pp. 425–428.

'**television documentary**', see: www.meanwoodpark.co.uk/insight/out-of-sight-the-documentary

'**studies of the genetics of intelligence**', for example, Shakeshaft N. *et al.* (2015), 'Thinking positively: the genetics of high intelligence', *Intelligence* 48, pp. 123–132.

'**Rushton**', Terry D. (2012), 'Leading Race "Scientist" Dies in Canada', *Salon*, 6 October.

EIGHT: Current Thinking

'**Orwell**', Carragee E. (2012), 'Penetrating neck injury: George Orwell is "struck by lightning"', *The Spine Journal* 12, pp. 769–770.

'**boosting mental performance**', Beveridge A. and Renvoize E. (1988), 'Electricity: a history of its use in the treatment of mental illness in Britain during the second half of the nineteenth century', *British Journal of Psychiatry* 153, pp. 157–162. And: Elliot P. (2014), 'Electricity and the brain: an historical evaluation', *The Stimulated Brain* (Elsevier), pp. 3–33.

'**mainstream science rediscovered**', Fox D. (2011), 'Brain Buzz', *Nature* 472, pp. 156–158.

'**US military**', Clark V. *et al.* (2012), 'TDCS guided using fMRI significantly accelerates learning to identify concealed objects', *Neuroimage* 59 (1), pp. 117–128.

'**contrary to popular belief**', Looi C. and Kadosh R. (2014), 'The use of transcranial direct current stimulation for cognitive enhancement',

Cognitive Enhancement: Pharmacologic, Environmental and Genetic Factors (Eds. Knafo S. and Venero C.) (Academic Press), p. 307.

'everyday life', Clark V. and Parasuraman R. (2014), 'Neuroenhancement: Enhancing brain and mind in health and in disease', *NeuroImage* 85 (3), pp. 889–894.

'driving', Santarnecchi E. *et al.* (2015), 'Enhancing cognition using transcranial electrical stimulation', *Current Opinion in Behavioural Sciences* 4, pp. 171–178.

'human corpse', Underwood E. (2016), 'Cadaver study casts doubts on how zapping brain may boost mood, relieve pain', *Science*, 20 April.

NINE: The Man Who Learned to Cry

'lasers', Barrett D. and Gonzalez-Lima F. (2013), 'Transcranial infrared laser stimulation produces beneficial cognitive and emotional effects in humans', *Neuroscience* 230, pp. 13–23.

'John Elder Robison', Quotes and details from interview with author, and from Robison J. (2016), *Switched On: A Memoir of Brain Change and Emotional Awakening* (Spiegel & Grau).

TEN: The Brain and Other Muscles

'multiple intelligences', Gardner H. (1983), *Frames of Mind: The Theory of Multiple Intelligences* (Basic Books)

'suggest the opposite', Visser B. *et al.* (2006), 'Beyond g: putting multiple intelligences theory to the test', *Intelligence* 34, pp. 487–502; and Visser B. *et al.* (2006), 'g and the measurement of multiple intelligences: a response to Gardner', *Intelligence* 34, pp. 507–510.

'emotional intelligence', Goleman D. (1996), *Emotional Intelligence* (Bloomsbury).

'elite football', Vestberg T. *et al.* (2012), 'Executive functions predict the success of top soccer players', *PLOS One* 7 (4), e34731.

'road cyclists', Okano A. *et al.* (2015), 'Brain stimulation modulates the autonomic nervous system, rating of perceived exertion and performance during maximal exercise', *British Journal of Sports Medicine* 49 (18), pp. 1213–1218.

'less committed', Vitor-Costa M. *et al.* (2015), 'Improving cycling performance: transcranial direct current stimulation increases time to exhaustion in cycling', *PLOS One* 10 (12), 16 December.

'tongue twister', Fiori V. *et al.* (2014), "'If two witches would watch two watches, which witch would watch which watch?" tDCS over the left frontal region modulates tongue twister repetition in healthy subjects', *Neuroscience* 256, pp. 195–200.

'Stubbeman', see: neuromodec.com/events/nyc-neuromodulation-conference-2015/abstracts

'learning to putt', Zhu F. *et al.* (2015), 'Cathodal transcranial direct current stimulation over left dorsolateral prefrontal cortex area promotes implicit motor learning in a golf putting task', *Brain Stimulation* 8 (4), pp. 784–786.

ELEVEN: The Little Girl Who Could Draw

'Nadia', Selfe L. (1977), *Nadia: A Case of Extraordinary Drawing Ability in an Autistic Child* (Academic Press).

'Treffert', from Treffert D. (2012), *Islands of Genius* (Jessica Kingsley Publishers), and from interview with the author.

'roll call', Treffert D. and Rebedew D. (2015), 'The savant syndrome registry: A preliminary report', *WMJ*, August, pp. 158–162.

'Pip Taylor', from interviews with the author.

'draw a person', Iqbal S. *et al.* (2013), 'Emotional indicators across Pakistani schizophrenic and normal individuals based on Draw a Person test', *Pakistan Journal of Social and Clinical Psychology* 11 (1), pp. 59–65.

'Bob', Simis M. *et al.* (2014), 'Transcranial direct current stimulation in de novo artistic ability after stroke', *Neuromodulation* 17 (5), pp. 497–501.

'provocative question', Nave O. *et al.* (2014), 'How much information should we drop to become intelligent?', *Applied Mathematics and Computation* 245, pp. 261–264.

'dementia patients', Miller B. and Hou C. (2004), 'Portraits of artists. Emergence of visual creativity in dementia', *JAMA* 61 (6), pp. 842–844.

'retired office worker', Takahata K. *et al.* (2014), 'Emergence of realism: enhanced visual artistry and high accuracy of visual numerosity representation after left prefrontal damage', *Neuropsychologia* 57, pp. 38–49.

'synaesthesia', Murray A. (2010), 'Can the existence of highly accessible concrete representations explain savant skills? Some insights from synaesthesia', *Medical Hypotheses* 74, pp. 1006–1012.

'Edinburgh study', Simner J. *et al.* (2009), 'A foundation for savantism?

Visuo-spatial synaesthetes present with cognitive benefits', *Cortex* 45, pp. 1246–1260.

TWELVE: The Genius Within

'shit got weird', http://www.xojane.com/it-happened-to-me/acquired-savant-syndrome

'foreign accent', for example, Verhoeven J. *et al.* (2013), 'Accent attribution in speakers with Foreign Accent Syndrome', *Journal of Communication Disorders* 46, pp. 156–168.

'near-death', Blanke O. *et al.* (2016), 'Leaving body and life behind: out-of-body and near-death experience', *The Neurology of Consciousness*, Second edition (Elsevier), pp. 323–347.

'epilepsy', Adachi N. *et al.* (2010), 'Two forms of déjà vu experiences in patients with epilepsy', *Epilepsy and Behaviour* 18, pp. 218–222.

'terrifying and distressing', both taken from http://www.epilepsy.com/connect/forums/new-epilepsycom/memorydream-flashes-after-seizures

'on demand', Moriarity J. *et al.* (2001), 'Human "memories" can be evoked by stimulation of the lateral temporal cortex after ipsilateral medial temporal lobe resection', *Journal of Neurology, Neurosurgery and Psychiatry* 71, pp. 549–551.

'odd side effects', Polak A. *et al.* (2013), 'Deep brain stimulation for OCD affects language: a case report', *Neurosurgery* 73 (5), E907–10.

'most dramatically', Tomasino B. *et al.* (2014), 'Involuntary switching into the native language induced by electrocortical stimulation of the superior temporal gyrus: a multimodal mapping study', *Neuropsychologia* 62, pp. 87–100.

'smart or dumb', Hesselmann G. and Moors P. (2015), 'Definitely maybe: can unconscious processes perform the same functions as conscious processes?', *Frontiers in Psychology* 6, pp. 584–560. And: Loftus E. and Klinger M. (1992), 'Is the unconscious smart or dumb?', *American Psychologist* 47 (6), pp. 761–765.

'that view was challenged', Sklar A. *et al.* (2012), 'Reading and doing arithmetic nonconsciously', *PNAS* 109 (48), pp. 19614–19619.

'University of Tulsa', Lewicki, P. *et al.* (1987), 'Unconscious acquisition of complex procedural knowledge', *Journal of Experimental Psychology: Learning, Memory, and Cognition* 13, pp. 523–530.

'Benjamin Langdon', Spitz H. (1995), 'Calendar counting Idiot Savants and the smart unconscious', *New Ideas in Psychology* 13 (2), pp. 167–182.

'savants make use of', Snyder A. (2009), 'Explaining and inducing savant skills: privileged access to lower level, less processed information', *Philosophical Transactions of the Royal Society* B 364, pp. 1399–1405.

THIRTEEN: The Happiest Man on Death Row

'William Sidis', Bates S. (2011), 'The prodigy and the press: William James Sidis, anti-intellectualism and standards of success', *Journalism and Mass Communication Quarterly* 88, pp. 374–397.

'Daily Telegraph', Kealey H. (2014), 'Why Do Geniuses Lack Common Sense?', *Telegraph*, 14 November.

'Clever sillies', Charlton B. (2009), 'Clever sillies: why high IQ people tend to be deficient in common sense', *Medical Hypotheses* 73, pp. 867–870.

'Joe Arridy', most taken from Perske R. (1995), *Deadly Innocence* (Abingdon Press).

'legal system to broaden', Greenspan S. *et al.* (2015), 'Intellectual disability is a condition not a number: ethics of IQ cut-offs in psychiatry, human services and law', *Ethics, Medicine and Public Health* 1, pp. 312–324.

'gullibility', Greenspan S. *et al.* (2001), 'Credulity and gullibility in people with developmental disorders: a framework for future research', *International Review of Research in Mental Retardation* 24, pp. 101–134.

'RQ', Stanovich K. and West R. (2014), 'The assessment of rational thinking: IQ ≠ RQ', *Teaching of Psychology* 41, pp. 265–271.

FOURTEEN: On the Brain Train

'Allan Snyder', Snyder A. *et al.* (2003), 'Savant-like skills exposed in normal people by suppressing the left fronto-temporal lobe', *Journal of Integrative Neuroscience* 2 (2), pp. 149–158.

'numerosity', Snyder A. *et al.* (2006), 'Savant-like numerosity skills revealed in normal people by magnetic pulses', *Perception* 35, pp. 837–845.

'solve a classic puzzle', Chi R. and Snyder A. (2012), 'Brain stimulation enables the solution of an inherently difficult problem', *Neuroscience Letters* 515 (2), pp. 121–124.

'expert chess players', Franke A. *et al.* (2017), 'Methylphenidate, modafinil, and caffeine for cognitive enhancement in chess: A double-blind, randomised controlled trial', *European Neuropsychopharmacology* 27 (3), pp. 248–260.

'Mozart effect', Pietschnig J. *et al.* (2010), 'Mozart effect – Shmozart effect: a meta-analysis', *Intelligence* 38, pp. 314–323.

'wasting their time', Underwood E. (2014), 'Neuroscientists speak out against brain game hype', *Science*, 22 October.

'asked viewers', Owen A. *et al.* (2010), 'Putting brain training to the test', *Nature* 465, pp. 775–778.

'silver-haired', Corbett A. (2015), 'The effect of an online cognitive training package in healthy older adults: an online randomized controlled trial', *JAMDA* 16, pp. 990–997.

'babies smarter', Lewin T. (2009), 'No Einstein in Your Crib? Get a Refund', *New York Times*, 23 October.

'Flynn effect', Flynn J. (2013), 'The Flynn effect and Flynn's paradox', *Intelligence* 41, pp. 851–857.

'tentative signs', Howard R. (2005), 'Objective evidence of rising population ability: a detailed examination of longitudinal chess data', *Personality and Individual Differences* 38, pp. 347–363.

'doomed to idiocy', Woodley M. *et al.* (2013), 'Were the Victorians cleverer than us? The decline in general intelligence estimated from a meta-analysis of the slowing of simple reaction time', *Intelligence* 41 (6), pp. 843–850.

'mathematics', Blair C. *et al.* (2005), 'Rising mean IQ: cognitive demands of mathematics education for young children, population exposure to formal schooling, and the neurobiology of the prefrontal cortex', *Intelligence* 33, pp. 93–106.

'smartest humans', Hsu S. (2014), 'Super-intelligent humans are coming', *Nautilus*, 16 October.

'genetic tweaks', Hsu S. (2014), 'On the genetic architecture of intelligence and other cognitive traits', *arXiv:1408.3421v2*, 30 August.

'pigs', Cyranoski D. (2015), 'Gene-edited micropigs to be sold as pets at Chinese institute', *Nature*, 29 January.

FIFTEEN: Faster, Stronger, Smarter
'IQ ranges', Kaufman A. (2009), *IQ Testing 101* (Springer), chapter five.